PACIFIC DEFENSE

ALSO BY KENT CALDER

The Eastasia Edge (with Roy Hofheinz, Jr.)
*Crisis and Compensation: Public Policy and Political
Stability in Japan*
*Strategic Capitalism: Private Business and Public
Purpose in Japanese Industrial Finance*

PACIFIC DEFENSE

ARMS, ENERGY, AND AMERICA'S FUTURE IN ASIA

KENT E. CALDER

William Morrow and Company, Inc.
New York

It is the policy of William Morrow and Company, Inc., and its imprints and affiliates, recognizing the importance of preserving what has been written, to print the books we publish on acid-free paper, and we exert our best efforts to that end.

Library of Congress Cataloging-in-Publication Data

Calder, Kent E.
    Pacific defense : arms, energy, and America's future in Asia /
  Kent E. Calder.
        p.   cm.
    Includes bibliographical references and index.
    ISBN 0-688-13738-5
    1. Arms race—East Asia.   2. Defense industries—East Asia.
  I. Title.
  UA832.5.C35   1996
  355'.03305—dc20   95–32617
                                                                      CIP

Printed in the United States of America

First Edition

1 2 3 4 5 6 7 8 9 10

BOOK DESIGN BY RENATO STANISIC

In memory of the Pacific past—
Mervyn S. Bennion and Edwin O. Reischauer

With hope for the Pacific future—
To Mari and Ryan Calder

PREFACE

Since the foundation of the American republic 220 years ago, this nation has looked eastward across the Atlantic, from whence the bulk of its forebears, both black and white, have come. The major figures in our country's consciousness—both friend and foe have been Atlantic, from the France of Lafayette and the Britain of George III to the Nazi Germany of Hitler, the Soviet Union of Stalin and his heirs, and the Indomitable England of Winston Churchill. Outside Europe, our major concern has been the Middle East.

Yet despite America's Atlantic preoccupation, many of our gravest challenges—in both security and economics, but preeminently at their interface—have come at the back door, from across the Pacific. From beyond its deceptively delicate, misty horizon—nearly inscrutable in the best of times—have emerged the confrontations that brought us World War II, Korea, and Vietnam. From the same direction has also come the vast bulk of our huge global trade deficit, recently averaging well over $100 billion annually.

Pacific Defense, for the United States, has two basic aspects. Most clearly there is protection against direct national security challenges to American forces—a repeat of the 1941 Japanese Pearl Harbor attack or the 1950 North Korean invasion across the DMZ, for instance. A second dimension of increasing practical importance, however, is dealing with destabilizing political rivalries and economic shortages within East Asia itself, which would otherwise threaten to generate more direct security threats to America over time.

Many of these are remarkably unknown to the general public, and unappreciated even by policymakers with Asian responsibilities.

This book is an attempt to present the impending long-term Pacific security and economic issues for the United States, in the dual aspects outlined above. It suggests that Pacific Defense has an increasingly important economic aspect, primarily in easing the resource insecurities that have plagued and driven East Asia since Pearl Harbor and before. Yet Pacific Defense also continues to have a central geostrategic dimension, despite the waning of the Cold War.

To be sure, we must continue to press nations across the Pacific for change in outmoded, mercantilist approaches to international trade. But our national agenda—even in its economic aspects—must be broader than one-dimensional confrontation over autos, mobile phones, flat glass, or insurance. The waning of the Cold War does *not* mean that economics is the only game in town.

The image of Pacific Defense begins symbolically for me, as indeed it does for many Americans, beneath a balmy sky, overlooking a tropical sea on a peaceful Sunday morning. At Pearl Harbor. The memories of the past have dimmed, and our deadly adversary that day in 1941 has become a potent ally. Yet the warning and the lesson—of the perilous costs we run from unpreparedness and naïveté about events beyond our shores—still remain.

Long before I ever went to Tokyo, Bangkok, or Beijing, my sentiments were captured by Captain Mervyn S. Bennion, commander of the battleship U.S.S. *West Virginia* in 1941, who was my father's cousin, and a legend around our household as I was growing up. In the early hours of December 7, 1941, Captain Bennion, standing on the bridge of his ship on Battleship Row next to the U.S.S. *Arizona*, was mortally wounded by fragments from a Japanese bomb. Refusing evacuation, he continued to command his burning, sinking ship to the last, and was posthumously awarded the Congressional Medal of Honor, one of sixteen awarded for heroism at Pearl Harbor on that dark day. Bennion's personal bravery, and the sudden brutality of the events that took his life, made a strong, still lingering impression on me.

If the memory of Mervyn Bennion indelibly taught me wariness about Pacific Defense, the memory of Edwin O. Reischauer, my Ph.D. thesis adviser, impressed on me the need for sensitivity in dealing with transpacific relations. Reischauer, born in Tokyo and then raised there during the turbulent interwar years, served in military intelligence during World War II and had little sympathy for Japanese militarism. Yet he clearly understood the indispensable basis of American action in the Pacific: two-way dialogue with Japan. And he knew that real communication could come only on an enduring basis of trust—such as he, and ultimately Jimmy Carter, established with Japanese prime minister Ohira Masayoshi or Ronald Reagan shared with Ohira's successor Nakasone Yasuhiro.

This book has been many years in the making. Its earliest intellectual roots lie in my doctoral work with Edwin Reischauer and Raymond Vernon at Harvard, which impressed on me the value of a long-term, micro-level, market-sensitive approach to transpacific problems. My years as first executive director of Harvard's Program on U.S.–Japan Relations, and with the Center for Strategic and International Studies in Washington, intensified my interest in an integrated economic and security approach to Pacific Defense.

Many colleagues, students, and friends—at Princeton, Harvard, CSIS, and elsewhere—have contributed greatly to the evolution of the ideas expressed here—so many that it is impossible to thank them all. David Abshire, James Auer, Henry Bienen, Toshiko Calder, Bill Clark, Ralph Cozza, Richard Fairbanks, Aaron Friedberg, Fukukawa Shinji, Robert Gilpin, Gerrit Gong, Richard Grant, Hoshino Shinyasu, Fred Iklé, Karl Jackson, Amos Jordan, Kannoh Tokio, Jane Khanna, Kosaka Masataka, Kusada Kaoru, Urban Lehner, Richard Missner, Wes Neff, Nishihiro Seiki, Joseph Nye, Gil Rozman, Saeki Kiichi, Satō Seizaburo, Robert Scalapino, Seiki Katsuo, Jim Shinn, Tanaka Akihiko, Usui Yoshinori, Barry Wain, John Waterbury, Lynn White, John Yochelson, and Donald Zagoria, together with several current government officials who unfortunately must remain nameless, have been among the most important. As throughout the book, Japanese names are presented in traditional form, with family name first. Edna Lloyd has done her usual efficient and remarkably patient job of preparing the manuscript. Hasegawa Mia, Hirano Eri, Alex Harney, Lee Howell, Chris Johnson, David Mitchell, Mark Meredith, Oyama Keiko, Ann Treistman, and Yanghee Woo have helped enormously with research assistance in the United States, while Izumi Koide and her staff at International House library have been most supportive in Japan. Norman Achilles, Solomon Karmel, and Kurt Tong have made insightful comments on various portions of the manuscript, while Henry Ferris has done a sensitive job of overall editing. No one, of course, agrees totally on such a controversial subject; responsibility for the views expressed here necessarily must be mine alone.

Princeton
January 1996

CONTENTS

PACIFIC DEFENSE

ARMS RACE ASIA?

Far more Americans have died in Asian battle over the past half century—from Saipan and Okinawa to Pork Chop Hill and Khe Sanh—than in any other part of the world. Yet we persist in our view of Asia as a prosperous, placid region. Quietly, however, a new danger zone has been rising, with global implications.

In the early 1970s the Middle East began arming itself with the windfall profits from oil price increases. Now Asia, with its newfound affluence as the world's lowest-cost and highest-quality manufacturer, is devoting much of its windfall profits to arms. Growth and emerging energy shortages are giving birth to dangerous new tensions. These build on ancient antagonisms, the residue of the Cold War, and the rising sophistication of local technology to spur a new strategic competition with disturbing implications for the new century now looming.

Pacific Defense for the United States in the Cold War years conventionally meant blunting Communist inroads by forward-deploying American forces in Asia, and supporting local stability and prosperity. It must also now confront a broader and more complex challenge: neutralizing the local tensions, insecurities, and shortages flowing from the explosive regional growth of recent years. These tensions could well spiral into buildups and new political orientations ultimately threatening nations far beyond Asia.

The old tools of Pacific Defense have been military, combined with liberal Asian access to U.S. markets. The new tools, to address redefined imperatives, may well have to vary also. A vigorous American Pacific presence—economic as well as military—remains crucial. Better access to Asian markets for foreign business is clearly needed. But hybrid, cooperative economic and security measures—on American initiative—to dampen regional rivalries, increase dialogue, and deal with shortages are crucial needs not yet adequately addressed. For as the tortured road to Pearl Harbor showed so clearly, Asia is at its most turbulent and dangerous to the broader world when it is insecure.

There is no shortage of Asian insecurities, both economic and political, to provoke regional tension in the present or the foreseeable future.

With the Cold War behind us, most nations around the world are slashing armaments. The United States, for example, cut its defense expenditures during the 1989–94 period by over $37.8 billion, or nearly 13 percent.[1] Russia, in the face of simultaneous economic and political collapse, cut even more sharply, by over 43 percent.[2] The Middle East has also seen major retrenchment, with Israel's share of GNP devoted to defense falling by half since 1985, and Saudi Arabia's by over a third, despite an enormous upsurge during the Gulf War.

East Asia is a sharp, disturbing exception to these positive trends in global hot spots. Almost every nation in the region save Vietnam, which already had 700,000 men under arms and one of the most formidable arsenals on earth, increased its defense spending by a third or more during the past decade. Several in Northeast Asia have done so in the last five years alone.

To be sure, East Asian economic interdependence is high, arms spending remains low relative to GNP, and local ministries of finance review defense budgets with a much more skeptical eye than has probably ever been true in the Middle East. There are some military categories (land forces) in which both force levels and budgets seem to be actually decreasing. Japan in 1994 recorded its lowest defense budget increase (0.8 percent) in thirty-four years[3] and announced plans to slash its ground force strength by 20 percent by decade's end. China's massive ground forces are also contracting.

Post–Cold War political changes are allowing at last some wary relaxation of bilateral tensions across the tangled old divides of the past. South Korea, in particular, has forged creative and rapidly deepening ties with both Russia and China since 1990. Vietnam is becoming reconciled with the United States, as GIs return to Danang in business uniform. Hanoi has even joined ASEAN, the Association of Southeast Asian Nations.

Yet despite some peaceful short-term signs, many longer-term tendencies are deeply troubling. Asian nations—from India and China to Taiwan and the two

Koreas—are rapidly developing new long-range strike capacities as legions of new missiles and aircraft become operational. They are building more and more weapons at home, creating vested long-run interests in further military expansion in years to come. And they are rapidly upgrading the level—and hence the destructiveness—of their domestic military technology, playing on the fierce post–Cold War rivalries of Western and Russian producers for the most buoyant remaining arms market on earth.[4]

The East Asian buildup is thus troubling not only for its pace—rapid in quantitative terms even by classic Middle Eastern standards of the 1960s and 1970s. Its technological level is also alarmingly high, with high-performance missiles proliferating, as is not surprising in a region famous for its electronic prowess. The buildup is fueled by the economic as well as strategic interests of local Asian arms producers, whose deadly exports of Silkworm and Rodong missiles provoke instability throughout the world. The arms race has important nuclear, chemical, and even biological dimensions not easy for outsiders to control, precisely because its key participants, especially China, are already so affluent, powerful, and technologically sophisticated in global terms.

The world has figured out how to deal with Saddam Hussein, although it was not easy. Monitoring of his arms programs is intensive, and sanctions cripple his ability to challenge the status quo. Yet coping with a slow-motion Asian arms race is much more frustrating, as U.S. dealings with North Korea have shown. Self-sufficiency and isolation give Kim Jong Il advantages that Saddam would envy. Though it lacks the explosive catalyst of ethnic conflict, the unsettling Asian buildup casts deep shadows on an otherwise bright Pacific future. In the post–Cold War world, it has eclipsed the Middle East as the most deadly long-term regional security challenge that the United States faces today.

Within East Asia, the trends are especially disturbing in the northeastern part of the region. Indeed, they surround economic superpower Japan, still hesitant and reactive with respect to security affairs, in a great "Arc of Crisis" stretching from southwest to northeast of Tokyo. Japan, China, Taiwan, and South Korea together spent significantly more on defense in the mid-1990s than did post-Soviet Russia, even by conservative estimates. Indeed, China's or Japan's defense spending alone could be comparable to Russia's, by some calculations, with actual Chinese spending as much as six times official pronouncements.[5] Much of Beijing's defense spending is reportedly camouflaged in civilian budget items and cross-subsidies from the profitable nonmilitary operations of defense enterprises, ranging from nightclubs to food production. Even the ice cream profits from the Baskin-Robbins 31 Flavors Beijing branch reportedly flow into the People's Liberation Army budget.

This massive $75-billion-plus Northeast Asian defense outlay represented an increase of over 50 percent for this volatile subregion over the 1990–95 period, and it marked an absolute scale of spending more than double what the United States devoted to its large Pacific forces. In addition, there looms North Korea, its spending difficult to quantify but the militance of its armed forces unquestioned.

Where geostrategic rivalries in Asia have already led us is starkly evident on the Korean peninsula. There more than a million men, including more than 37,000 Americans, confront one another across a DMZ six miles wide. More than eight thousand North Korean recoilless artillery pieces, dug into blast-hardened underground concrete silos only twenty-five miles from Seoul, threaten credibly to turn South Korea's capital of twelve million people into a "sea of fire" at any time. And unanswered questions on the North's nuclear capacity leave open the prospect, one hopes declining, that the flames could be radioactive as well. In response, with one eye turned also to perceived geostrategic uncertainties in China, Japan, and the oil routes to the Persian Gulf, South Korean defense spending has soared over the past decade at the most rapid rate in Asia.

The lingering specter of the Cold War past is also clear at Vladivostok, only four hundred windswept miles farther along the Northeast Asian Arc of Crisis, at the edge of the forbidding Siberian landmass. There, in a secluded bay surrounded by one of the largest stockpiles of tactical nuclear weapons in the world, with spiraling surveillance radar on the mountaintops, the heart of Russia's Pacific fleet lies at anchor. Invisible but also a formidable part of commanding admiral Igor Khmelnov's realm are nearly twenty strategic nuclear submarines circling beneath the icy Sea of Okhotsk, with their ICBMs traditionally targeted on Seattle, Portland, and San Francisco, and the durability of a post–Cold War change in such practices still uncertain.

Coupled with these persisting images of the past are stark new specters of the future. At the southern apex of the Arc of Crisis thousands of hard-hatted Taiwanese workmen have been feverishly at work, hollowing out caves in the jagged granite mountains of Taiwan's southeast coast at Chiashan, near the farthest spot in Taiwan from the Chinese mainland. There the Nationalists hope to shelter against the threat of Chinese air strikes the 150 F-16 fighter-bombers that George Bush sold to Taiwan in a $6 billion 1992 election-eve deal that carried for him the state of Texas. The base has its own power generators, a microwave landing system to allow multiple landings and takeoffs, and several months' supply of food, fuel, and military stores. It is a veritable land-based aircraft carrier, for confronting an uncertain future.

Simultaneously, China buys fighters from Russia, while fostering the omi-

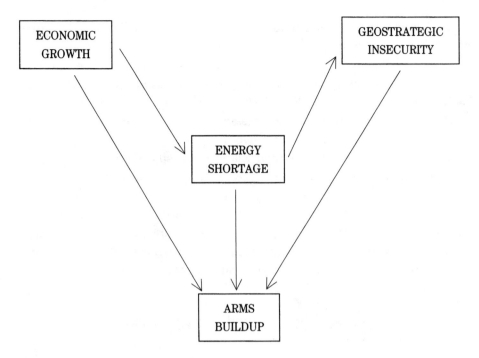

Figure 1-1

nous embryo of its first blue-water navy in five hundred years. The sea-lanes of the China Seas and the Indian Ocean, not to mention the vicinity of Taiwan, are all obvious fields of maneuver. As China considers adding aircraft carriers to the strategic waterways of Asia, in the wake of intermittent Taiwan Strait missile confrontations, the unease in Tokyo is palpable.

The shadows of this dusk-tinged future, pregnant with implications for Pacific Defense, paradoxically flow from precisely the affluence and explosive growth that make Asia's prospects appear so bright at first glance. The key to the problem, in its starkest terms, is the deadly quadrangle presented in Figure 1-1: growth, energy, geostrategic insecurity, and armament.

Energy: Asia's Achilles' Heel

Across the entire fifteen-hundred-mile sweep of the Japanese archipelago, there is not now—and has never been—a single major developed oil field. Ninety-nine percent of total supply comes from imports, mostly from the distant Persian Gulf. The rest of Northeast Asia—Taiwan, the Russian maritime provinces, Korea (both North and South), and North China as the Taching fields deplete—is in similar straits. Even the abundant North Sumatran fields of Indonesia—which provoked

Japan's armies, in their thirst for fuel, toward war in 1941—may well be exhausted by the turn of the century.

As Asia grows, its demand for oil soars still faster, fueled by the consumer revolution sweeping the region. One million cars annually are sold now in the seven ASEAN nations, as well as 300,000 in China—more than double levels of just five years ago. These numbers are expected to double again in Southeast Asia—and could rise tenfold in China—by the year 2000.[6] China's "People's Car" project may send regional oil demand sharply higher still in the following decade. Petrochemical industries, and the related use of plastics and fertilizers, are surging regionwide, spawning further demand for oil.

Asia's thirst for energy goes far beyond petroleum. As its economy grows explosively, the region ravenously consumes ever larger amounts of coal and natural gas. Yet those too are in short supply. The nearly inevitable result: energy crisis in a region representing a quarter and more of the global economy, and an ever larger share of its energy demand.

The Nuclear Shadow over Asia

How will Asia slake its thirst for energy? One option we have already seen at Chernobyl, Three Mile Island, and most recently Yongbyon, North Korea: nuclear energy. The OECD anticipates that Asia, primarily energy-short Northeast Asia, will introduce forty gigawatts of new nuclear capacity between 1992 and 2010— 48 percent of the global total.[7] Many of these new Asian nuclear plants, like North Korea's Yongbyon, will generate dangerous amounts of plutonium, the raw material for nuclear weapons. Japan's breeder reactor program, based on reactors that produce more fuel than they consume, could ultimately break that nation's crippling dependence on the outside world for energy by the middle of the next century. But Japan could in the process amass close to one hundred tons of plutonium—more than is contained in all the current nuclear weapons of the United States and Russia combined—even should it not explicitly elect to go nuclear. China and South Korea are also seriously considering fast-breeder programs. North Korea's Yongbyon, for all our trials in dealing with it, will prove to be just the tip of the iceberg.

Energy demand will thus steadily propel Asia to the threshold of nuclear proliferation. And political rivalries, especially in the volatile Northeast Asian Arc of Crisis surrounding Japan, where three current nuclear powers (Russia, China, and the United States) converge around an unstable, divided, and heavily armed Korean peninsula, could well push it across that ominous threshold. We have already seen the prospective dangers of the nuclear Pandora's box in the bitter political struggle over North Korea's reactor program. Any number of ongoing rivalries, rooted deeply in history, could force Northeast Asian states to acquire nuclear weapons: the Taiwan Strait controversy, tensions within Korea, or Japanese tensions with continental powers in the wake of Korean reunification, to cite a few possibilities. The ambiguous October 1994 resolution of North Korea's nuclear status does not defuse this unsettling prospect of further weapons-related turmoil.

Bottlenecks in Asia's Vulnerable Oil Lifeline

Asia's energy shortages give rise to security tensions in another crucial way: they force ever-rising imports from outside the region, especially from the far-distant Persian Gulf. Wellhead production costs of $1.50 or less per barrel in the Gulf—where oil almost oozes from the sand—make it the natural place for oil-short Asia to slake its thirst. By the year 2010, dozens of supertankers a month will be departing Saudi wellheads like Ras Tanurah for Pusan, Yokohoma, and increasingly Shanghai; over fifteen million barrels daily (nearly 20 percent of global consumption) could well be flowing from the Gulf as a whole to the Far East.[8]

As the massive supertankers wind their lumbering way eastward, their course could be a precarious one, should the American commitment in the area weaken. From the broad waters of the Indian Ocean, they must run a gauntlet of proliferating Indian, Malaysian, Indonesian, and Singaporean naval installations and listening posts, many still under construction, surrounding approaches to the six-mile-wide Strait of Malacca. Alternatively the largest tankers, which cannot safely navigate the strait, must steam around Java and through the Lombok and Makassar straits. These waterways, the best deepwater routes to Northeast Asia, are dominated by an Indonesian navy recently reinforced by purchase of the entire East German preunification fleet.

China: A Powerful New Oil Importer

Northeast from Singapore, through the waters where Japanese attack planes sunk the battleship *Prince of Wales* in early 1942, East Asia's shipping lifeline passes other choke points: the Spratly and the Paracel islands of the South China Sea. The Paracels, captured by Chinese marines in 1974 while both Vietnams were diverted just before the fall of Saigon, already feature a long-range airstrip at which Su-27 fighters from China's mainland routinely land. Five hundred miles to the southeast across the sea-lanes, the Spratlys, known as the "Dangerous

Ground" by ancient mariners because of treacherous shoals, lie engulfed in a thicket of conflicting assertions by six rival claimants.

The shadow of China looms large in these strategic southern waters. Potentially huge amounts of subterranean oil and gas compound their attraction, for both China and others. A major oil strike has already been made off the nearby Filipino island of Palawan, which prompted a Chinese land grab only 170 kilometers away in early 1995. A massive natural-gas find—the Natuna field off Indonesia—has also recently been made and has likewise rapidly found itself within Chinese waters on Chinese maps. Some even claim another Saudi Arabia lies beneath these placid seas, although a welter of conflicting claims makes this potentially dangerous to verify.

The treasure fleets of Admiral Zheng He passed this way over five hundred years ago, and the Ming dynasty regarded these waters as a virtual internal sea of the Chinese empire. Today China quietly prepares to enforce its historic claims with new bases on Hainan and at Zhanjiang in southern Kwangsi, and the impending leverage of in-flight refueling capacity for its MiG-29s and Su-27s. With such enhancements they can fly confidently on patrol to the deepest reaches of the South China Sea, where China's most ambitious claims expire virtually within sight of Indonesia.

More and more supertankers from the Gulf unload their precious cargo at Chinese ports, a trend likely to accelerate in coming years. Since November 1993, China has been a net oil importer; double-digit growth in its energy-deficient southern provinces steadily swells China's shortfall despite repeated Beijing efforts to suppress it. As Chinese oil imports steadily rise, defending the fragile sea-lanes to the far-off Persian Gulf becomes a new security imperative for the PLA Navy. So does enforcing claims to prospective offshore oil-drilling rights in the disputed yet prospectively oil-rich East and South China seas, directly adjoining China's most energy-deficient regions. Not surprisingly, China has been training carrier pilots, acquiring appropriate aircraft, and preparing to build its first two aircraft carriers, suggesting plans for a blue-water navy that would fundamentally restructure the Asian strategic equation.

The problem for Asian stability, growing with each barrel of Chinese oil imports, is now clear. It is the danger that China's attempts to safeguard its oil supply lanes and defend its historical "sovereignty" in adjacent seas poses for other nations of Asia, especially for Japan. China claims 80 percent of the South China Sea as territorial water; 70 percent of Japan's oil supplies pass that way. Figures are similar for South Korea, and indeed Taiwan. All three Northeast Asian nations will feel an impulse to resist or counterbalance both rising Chinese naval power and the tightening grip of lesser states over the many bottlenecks across their vital oil lifeline to the Persian Gulf.

The Shadows of Reunification

Apart from prospective energy shortage and its consequences, the other large shadow over Asia today is the future of its two divided nations: China and Korea. With the Cold War waning, much of the global geostrategic rationale for division has disappeared. Economic and personal ties are beginning to deepen, especially across the Taiwan Strait. Taiwan now has well over $20 billion invested in China, which absorbs fully 10 percent of its total exports. Since mid-1993, ordinary letters have been passing back and forth across the strait. It is now possible for any but high Beijing officials to spend up to two months in Taiwan, while Taiwanese tourists in the hundreds of thousands visit China. Even in Korea, economic ties between North and South are stirring.

Yet amid this deepening interdependence—which makes unification among kindred Chinese and Korean peoples seem ever more plausible—dangerous cross-pressures are also escalating. The arms buildups on both sides of the DMZ and the Taiwan Strait have already been alluded to. The missile rattling and ostentatious military exercises are well known. As both sides in divided China and Korea come ever closer—with unity ever more visible on the horizon—both sides grow ever more grimly determined not to be seduced by their interdependence. Both maneuver for maximum leverage for themselves. Arms, sadly, are the ultimate equalizer, and brinkmanship is basic to their political utility.

The dual uncertainties of the Chinese and Korean futures are especially worrying to Japan, which lies next to both of them and is much less heavily armed than either. Taiwan is not only a mere half hour flying time from Japan's southern bastion of Okinawa, but lies directly across Japan's energy lifeline to the Gulf. Korea's gray outline lies nearly within sight of Japan at Shimonoseki, looming over its approach to the Asian continent.

Meiji Japan valued the strategic location of Taiwan enough to conquer it within thirty years of the Restoration itself. Meiji *genro* Itō Hirobumi was so obsessed with Korea, for him "a dagger pointed at the heart of Japan," that he inspired its occupation only fifteen years later and became its first resident general. Times obviously have changed, but Japan has not been exposed to the geopolitical reality of a Taiwan or Korea in truly hostile hands for nearly a century. In Taiwan's case it has been more.

The interactive relationships between unification moves in China and Korea, on the one hand, and a Japanese response will be highly complex, and related to Japan's broader alliance ties. Since unification is an internal matter for those two nations, it is difficult for Japan to express itself directly, and it has not done so. Yet Tokyo is clearly concerned, especially as it sees arms spending within the two divided nations escalating, even as their internal economic interdependence also rises.

China and Korea, with some historical reason, have never trusted Japan's intentions. The shadows of unification, ironically, deepen their suspicions as well. They too can see the challenges that their changed geo-economic standing could pose to Japan. And many of them are not sanguine about its response, openly projecting Japanese rearmament in coming years, despite the clear constraints within Japan itself.

America's Role?

A mere five hundred miles—just twenty minutes as the F-16 flies—from the naval dry docks of Shanghai, now relentlessly building the new Chinese fleet, lie the rocky volcanic reefs of Okinawa. To the uninitiated these islands, with their pristine white and black sand, so striking against the aquamarine of the East China Sea, may seem a simple isolated paradise—a kind of Oriental Hawaii. Yet there is no more strategic or evocative location on earth.

Thirteen thousand American Marines, together with 240,000 Japanese (among them more than 100,000 local civilians), died here during the bloody spring and early summer of 1945, in the only military engagement with a foreign foe ever fought on native Japanese soil. Today the sprawling Kadena Air Base, where the F-16s lie deployed in reinforced concrete revetments, remains a crucial anchor for the American strategic presence in the western Pacific and a security guarantee for the lightly armed economic giant Japan.

Kadena's fighter-bombers lie poised less than 500 miles from the Taiwan Strait, 520 miles from Shanghai, 900 miles from the DMZ in Korea, and only slightly more from Hanoi, Beijing, the shipping lanes of the South China Sea, and the headquarters of the Russian Pacific fleet at Vladivostok. Across the volatile lands of Northeast Asia's Arc of Crisis, they are crucial stabilizers. The U.S. Marine base at Camp Henry, a few minutes by jeep from Kadena, threw seven thousand men into the breach against Saddam Hussein at the outset of Desert Shield, and remains the only brigade of U.S. rapid-deployment troops stationed in the western Pacific.

Yet there is a faded, ominously archaic character to the strategic Okinawan outposts of America, reminiscent of the Panama Canal Zone a generation ago, that contrasts starkly to the rising specters of future Asian insecurity that we have seen. The officers' homes—incredibly spacious and well-watered by Japanese standards—are aging, and resented by locals living cramped on one-quarter the visitors' quota of space. The upkeep is getting costly, even with rising Japanese government subsidies to pay the hired help. American Fords and Chevys dot the highways, but they are mostly older models, brought from stateside by the servicemen at the expense of Uncle Sam. The Japanese, spurning used Fords, drive newer Toyotas and Nissans.

The U.S. Navy receiving depot still remains in the heart of Naha Port, faintly exotic after the Gulf War with its shiploads of half-tracks in brown-beige desert camouflage paint, returned from the Middle East. But the Navy's lumbering heavy vehicles generate more and more resentment as they tie up spiraling commuter traffic at rush hour, fueled by auto registrations that have more than tripled since Okinawa's formal reversion from the United States to Japan a generation ago, despite recent promises to relocate sometime within a decade. Senseless, sensational crimes, like the brutal rape of a twelve-year-old Okinowan girl by three GIs in the fall of 1995, also increasingly inflame local frustrations.

Asian Insecurity and Global Turmoil

Asia, with all its uncertainties, now casts a larger and larger shadow on the world scene. Its economy already makes up a third of the global market and 41 percent of global bank reserves, up from 17 percent in 1980.[9] But with half of all the people on earth, high savings, ever more sophisticated technology, and explosive, often double-digit growth rates across much of its periphery, the region seems destined for an ever greater share of global product. Japan and Greater China[10] alone hold two-thirds of the foreign exchange reserves on earth.

Developments in Asia, reflecting Asian economic power, increasingly set prices in the United States, and indeed around the world. Our markets for energy, food, and capital all lie hostage to Asian developments, in this age of interdependence. Even shrimp and soybean prices, not to mention Treasury bills, fluctuate with the yen's undulations. Yet our stabilizing security involvement in Asia, which inhibits local political rivalries and the struggle for energy, is ever more tenuous, as Okinawa's trends, including large recent demonstrations, so clearly tell us.

The deepening regionalist shadows on the world economy, and perennial protectionism, make projecting the status quo confidently ever forward an exercise in wishful thinking. Will Americans forever stand guard for the world's richest region far across the Pacific, even as their economic interests turn closer to home?

Will Asians themselves want to pay the cost, and endure the frustrations, of America's indefinite presence?

Asia, like America, is developing other options. Within, it confronts the most buoyant potential future market on earth. The Achilles' heel of energy, to be sure, remains. Yet there is the Middle East, including Saddam's Iraq and the Iran of the ayatollahs, with two-thirds of global oil reserves to slake even the most ravenous thirst. If copious dollars from Asia's mounting global surpluses do not suffice, there is always technology—even missile or nuclear—that Beijing or Pyongyang may willingly supply. Energy thus deepens the prospect of a renegade Islamic-Confucian embrace.[11]

Where will all this lead? The tides of history, unless someone turns them, are flowing out of Asia for the United States. Asia, to paraphrase Mao Zedong atop the Gate of Heavenly Peace, as he declared his People's Republic in 1949, is standing up. Yet Asia remains turbulent, with energy needs, nationalist antagonisms, post–Cold War uncertainties, and a proclivity for combustible arms competition fueled by double-digit growth. These strivings will render it a dangerously volatile region in coming years, in the absence of strong American involvement. And Asia's volatility amid affluence will doubtless cast deep shadows over the new twenty-first century now struggling to be born, making Pacific Defense of concern for all the world.

THE NORTHEAST ASIAN ARC OF CRISIS

Less than two hours from peaceful Tokyo by supersonic aircraft, or a matter of minutes as a Rodong missile flies, lies the DMZ in Korea, a focal point of Pacific Defense for nearly half a century. Along its troubled contours more than one million men under arms confront one another in a deployment area no more than forty miles wide. From the DMZ it is less than three minutes by MiG-29 to the Blue House, home of South Korea's president, in Korea's capital city of Seoul. That ramshackle metropolis, which North Korea periodically threatens to turn into a "sea of fire," is home to twelve million people, or nearly one-quarter of the entire South Korean population.

Throughout Seoul, in sharp contrast to Tokyo, the military is omnipresent, with major city streets strategically widened to serve as tactical military airfields. With good reason: should the North attack, Korea's capital would immediately become a battlefield, as it incessantly was in 1950 and 1951. Although such an onslaught would be suicide for Pyongyang, military pressure is the only form of leverage it has, given its desperately weak, declining economy. And important groups within the North have little to lose from confrontation.

Seoul and the DMZ are but the apex—the prospective epicenter, in the grimly appropriate terminology of earthquake science—of a much broader realm of in-

CHAPTER **2**

stability, stretching from the Taiwan Strait across North China and Korea to the Russian Far East. This Northeast Asian Arc of Crisis subtends the territory and deployed forces of three nuclear powers and two divided nations, still estranged by a Cold War not yet subsided. It encircles Japan from the southwest to the northeast; its future has fateful implications for the geopolitical orientation of that economic superpower, which is far less predetermined in its long-standing near-pacifistic course than often thought.

Much as a Middle Eastern Arc of Crisis from the Horn of Africa to the Persian Gulf surrounds the Arabian Peninsula and fatefully affects its political stance, so Arabia's Northeast Asian analogue can crucially shape the fate of Japan, and in turn the future of Pacific Defense. With a sixth of global GNP and state-of-the-art defense-industrial technology, yet complex domestic political constraints on military deployment, Japan has enormous latent military potential that if fully mobilized could reconfigure the face of world politics. Just as the policies of the Carter, Reagan, and Bush administrations were inevitably concerned with turmoil on the fringes of the Persian Gulf and its vast oil reserves, from the days of Khomeini's 1979 revolution to the Gulf War of 1991, so in coming years must American policy concern itself systematically with the volatile, yet high-growth, lands of Northeast Asia.

A Recurring History of Turbulence

Northeast Asia, with its lack of unifying institutions and human networks, has been an Arc of Crisis many times before. Indeed, its heritage of turmoil dates back more than seven hundred years. Kublai Khan's invasion fleets twice tried reaching Japan from Korea in the late thirteenth century, only to be forestalled by a *kamikaze*, or divine wind, that decimated them on the eve of departure. This gave birth to a myth of Japanese invincibility that sustained countless suicide missions in the emperor's name across the last dark, bitter days of World War II.

In 1592, the megalomaniacal upstart Hideyoshi surged in the opposite direction across the Strait of Tsushima with nine brigades, to lay waste to Kyongjyu and other venerable Korean cultural centers, in his grandiose effort to conquer all of the known world for Japan. Japan moved in once again after the forcible opening of Korea's "Hermit Kingdom" to the outside world in the 1870s. Tokyo established a powerful economic and military presence that eventually led in 1910 to formal annexation, despite the 1909 assassination of the great Prince Itō Hirobumi himself, who had just served three years as imperial resident general in Seoul, at the hands of a fervent Korean nationalist.

The bitter heritage of Japan's subsequent imperial presence in Korea darkens

Figure 2-1

NORTHEAST ASIAN ARC OF CRISIS

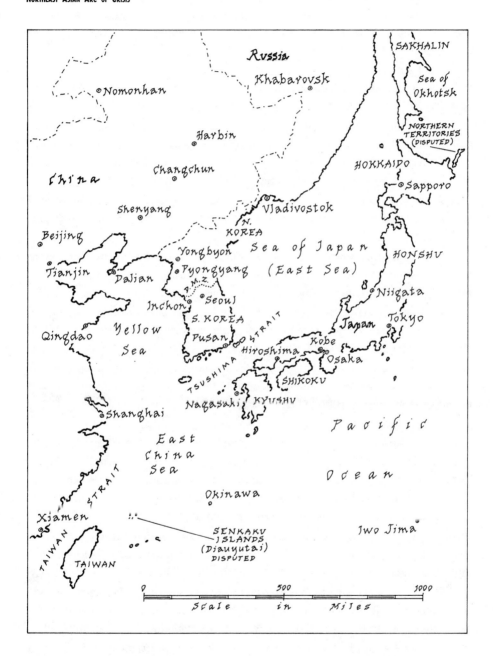

relationships across the Japan Sea (for Koreans, emphatically the East Sea) to this very day. Korean peasants by the hundreds of thousands were dispossessed to make room for Japan's own rural poor. In the schools of the peninsula even the speaking of Korean was forbidden, and all students were given Japanese names. Korean men were conscripted for forced labor battalions and the lowest, most dangerous ranks of the Japanese army, while thousands of Korean women were also kidnapped as forced "comfort women" to serve Japanese troops at the front.

Korean memories of the Japanese occupation did not die easily. Until 1965, two decades after World War II, South Korea steadfastly refused even to recognize Japan, despite American encouragement and the obvious economic attractions. To this day the broadcast of Japanese-language songs by the local mass media is forbidden. The "comfort women" issue continues to provoke large, emotional demonstrations against Japan, half a century after the war's tumultuous formal end.

With no larger framework existing to curb the ambitions of the many great powers jostling one another in its vicinity, Manchuria has, like Korea, been an unstable arena of conflict throughout most of this century. Until 1904 the Russians and the Japanese struggled for control of the Manchurian railways; that conflict was violently resolved by the Japanese surprise attack on czarist Russia's Pacific fleet at Port Arthur, followed by Tōgō's Tsushima naval victory of 1905. Yet a less sanguine but more protracted struggle for economic supremacy continued between Japanese and Western interests until 1931. At that point Japan invaded and established the puppet state of Manchukuo, with the last Manchu emperor, Henry Pu Yi, at its head.

During the late 1930s, Japan and the Soviet Union waged a protracted, bloody, but undeclared war along the Manchurian frontier, culminating in the massive yet virtually unreported battle of Nomonhan during the summer of 1939, which left more than a full division of imperial troops dead or missing. This bloodletting for weeks sent casualties by the thousands trickling back under cover of night to the hospitals and mortuaries of Japanese-occupied Harbin. It convinced the Imperial Army, with Tōjo Hideki at its ultimate head, to strike south against the East Indies and Pearl Harbor rather than north toward Siberia in the fateful winter of 1941.[1]

The weighty heritage of World War II itself in the region is almost beyond reckoning. For Chinese, Japanese, and Koreans, it lasted much longer than for Westerners—from 1931, the Japanese invasion of Manchuria, or from 1937, when the Kwantung Army stormed across Tianjin's Marco Polo Bridge into the heart of China itself. In eight bitter years of the Sino-Japanese War, over twenty million Chinese died, together with many more Japanese (and Korean conscripts) than fell at the hands of the United States from Pearl Harbor through Hiroshima and Nagasaki.

During the final days of World War II, the Northeast Asian Arc of Crisis witnessed brutal and widespread Red Army attacks against both Japanese Imperial Army forces and Japanese civilians. This Soviet onslaught took 300,000 formal prisoners and led to atrocities from Sakhalin to Manchuria, many of them well after Japan's formal surrender. The events of this period also gave birth to the so-called Northern Territories problem: the Soviet Union moved into parts of traditionally Japanese territory that were not clearly covered by the Yalta agreement sanctioning Soviet recovery of the Kuril Islands, and it did so even after the August 15, 1945, cease-fire ostensibly ending the war. Taken together, the callous treatment of Japanese prisoners and the postwar seizure of disputed territory left a deeply embittered heritage. That has continued to poison Russo-Japanese relations right down to the present day.

The guns of the Pacific war were only barely silent when Northeast Asia exploded again, in the flames of the Chinese Revolution. To be sure, the Kuomintang of Chiang Kai-shek and the Chinese Communist Party of first Li Li-san and then Mao Zedong had been playing a lethal cat-and-mouse game ever since 1927, when Chiang turned on his former allies and tried to exterminate the Communists. But the epic struggle came to a head in the wake of Japanese defeat. After savage battles in Manchuria during 1948, the People's Liberation Army swept into Beijing in January 1949 and surged across the Yangtze by summer. Chiang Kai-shek retreated to Taiwan, beginning an uneasy truce across the strait that remains both unresolved and unamenable to outside mediation.

Less than a year later, the Northeast Asian Arc of Crisis once again exploded in the flames of war as North Korean T-38 tanks rolled across the 38th parallel into the South. Three long years of conflict ensued, embroiling the United States, China, and fifteen other nations, crippling the entire Korean peninsula, and killing more than three million of its inhabitants. The heritage of bitterness has continued to this day, with more than a million men still confronting one another across the DMZ. They face a fragile peace governed only by armistice, even as the Cold War becomes a mere memory elsewhere in the world.

Why Northeast Asia Is More Troubled

Given its turbulent history of bitter and continuing conflict among great powers, it is not surprising that the Northeast Asian Arc of Crisis remains heavily and dangerously armed to this day. As indicated in Table 2-1, the Korean peninsula alone has 1.8 million men under arms—more than either the United States or the

Table 2-1
A More Militarized Northeast Asia

| | Northeast Asia | | | | | Southeast Asia | | | | | |
	China	North Korea	South Korea	Taiwan	Japan	Vietnam	Indonesia	Thailand	Malaysia	Philippines	Singapore
GNP	509	21	380	240	4,592	19	155	132	70	64	61
Population	1,201	24	45	21	125	73	193	60	20	69	3
Current account surplus	+6.4	na	−4.5	+8.2	+117.6	NA	−3.7	−6.7	−1.6	−1.0	+2.9
Defense budget	7.5	2.2	14.4	11.3	53.8	0.9	2.6	4.0	2.4	1.0	4.0
Armed forces	2.93m	1.128m	633,000	376,000	239,500	572,000	274,500	259,000	114,500	106,500	53,900
Nuclear capacity	yes	probably nuclear	near nuclear	near nuclear	near nuclear	no	no	no	no	no	no
Major local defense industry	yes	yes	yes	yes	yes (no export)	no	some	no	no	no	no

Source: IISS, *The Military Balance,* 1995–96 ed. Far Eastern Economic Review, *Asia Yearbook,* 1995 ed.
Notes: Defense budget figures for fiscal 1995, except the Taiwan figure, which is for 1994. The figure for China is the official amount; the IISS estimate for 1994 is $28.5 billion. Economic figures are generally for 1994, except current account figures, which are for 1992. Economic figures are $ billions. Population figures are in millions.

former Soviet Union. In total, more than five million troops are mobilized in the four states around Japan's periphery—almost 20 percent more than in all the armed forces of NATO.[2]

Looming over all Northeast Asia, as suggested in Chapter 1, are the shadows of reunification. Today China and Korea are armed but divided, while Japan rests with only limited deployments beneath the American security umbrella. How that triangle will evolve across the coming decade, as pressures for reunification proceed, is very complex and difficult to predict. Yet surely we know that the stakes will be high and the reactions sharp among all the national players involved.

Political-military tensions are further heightened in Northeast Asia by the nuclear equation. China is a major nuclear power, while Russia and the United States have substantial nuclear forces in close proximity. Vladivostok, home of the Russian Pacific fleet, reportedly houses one of the largest tactical nuclear stockpiles in the world, while the nearby Sea of Okhotsk is home to 40 percent of Russia's strategic seaborne nuclear deterrent. North Korea is widely believed to have several crude nuclear devices, while South Korea and Taiwan are both close to nuclear status.

Reflecting their substantial arsenals, the nations of the Northeast Asian Arc of Crisis maintain substantial defense expenditures—well over $50 billion in 1995. Combined with those of Japan, Northeast Asian expenditures are significantly larger than those of the former Soviet Union.

Reflecting its heavy military spending, Northeast Asia is also developing a qualitatively impressive arsenal. All the nations of the region have advanced fighter aircraft and are devoting major efforts to deploying advanced tactical missiles, many of them foreign-made. According to the Stockholm Institute of Peace Research, 73 percent of all East Asia's arms imports over the past decade have been undertaken by China, Japan, Taiwan, and the two Koreas.[3]

As is suggested later in more detail, Northeast Asian military expenditures have been rising rapidly since the end of the Cold War. And the economic scale of the nations in the region is such that they can continue to sustain major increases in the future. Another factor enhancing this prospect is the existence of active local defense industries, with sophisticated technology, in every nation of the region.

As Table 2-1 suggests, the situation is different in Southeast Asia along virtually every dimension. The "shadow of reunification," which continues to bedevil international politics in Northeast Asia, has been resolved farther south. Vietnam's government in Hanoi, the beneficiary of that forceful resolution more than twenty years ago, retains a powerful, battle-hardened military. But it has neither nuclear weapons nor other attributes of major political-economic power. And none of the other nations in Southeast Asia comes close to Vietnam in political-military capabilities or potential. Furthermore, none of the Southeast Asian nations has substantial local defense industry, in sharp contrast to patterns farther north.

Despite Northeast Asia's multiple security uncertainties, it has no multilateral security framework whatsoever, virtually alone among the world's major subregions. In contrast to this unstructured situation, Southeast Asia has a relatively rich framework of cooperative organizations binding its members. Most conspicuously, there is ASEAN, founded in 1967. The ASEAN Post-Ministerial Conference, founded in 1979, at which security issues are discussed with South Korea, Japan, and a range of Western nations including the United States, is also very active.

Economically, Northeast Asian growth tends to be volatile, because of the salience in local political economies of large industrial conglomerates (*keiretsu* in Japanese and *chaebol* in Korean). These large groups integrate finance and marketing closely with industrial production. Supported by a panoply of banks and insurance companies that will flexibly provide them credit and by general trading companies that will aggressively market their output, manufacturers tend to adopt high-volume, market-share-oriented strategies. Their economies also have the broader political institutions to promote high growth, including sympathetic bureaucrats; indeed, like a Ferrari's engine, their high-geared economies operate best at high speed. Yet as in racing also, the crashes and shocks, when they occur, can be monumental.

Southeast Asian economies are less highly geared. Family firms, many of them overseas-Chinese-owned, are dominant, and most of them are small. In contrast to the Japanese and Koreans of the Northeast, most have neither the financial nor the marketing structure to confidently support large-scale, highly leveraged growth. Southeast Asian economies, however, are thus spared the roller-coaster oscillations of Northeast Asia and glide along on a more even keel.

To say that growth may be smoother in Southeast Asia is not to deny the shadows of uncertainty that nevertheless threaten to complicate it. Most important among them are the lingering dangers of conflict over energy, a prospectively vital commodity to Asia's future, which have been dramatically manifest in a series of Vietnamese and recently Philippine confrontations with China. Indeed, a substantial share of Asia's prospective energy reserves may well lie in the hotly contested waters of the South China Sea, where these conflicts have occurred. But with much smaller economies and lower levels of armament in Southeast than in Northeast Asia—not to mention the absence of nuclear weapons—the potential for large-scale, highly destructive violence among local powers of the Southeast Asian region is relatively limited.

The Special Role of Japan

Japan, for a range of historical and economic reasons, relates somewhat differently to Northeast than to Southeast Asia. Since it is both the largest economic power of the region, with roughly half of total Asian regional GNP, and the largest direct investor in other nations, this contrasting relationship to the two areas has major implications for regional development. Broadly speaking, the heritage of history between Japan and Northeast Asia is a bitter one. To make matters worse, the nations of that subregion find themselves vigorous economic competitors. In contrast, the heritage of history between Japan and Southeast Asia is less contentious, while economic complementarities contribute to a more positive overall relationship than that prevailing in Northeast Asia.

In contrast to the brutal colonial rule and large-scale wartime casualties that prevailed on both sides in Korea, Taiwan, and China, as noted above, Japan's historical ties with much of Southeast Asia have been more positive. To be sure, Japanese wartime relations with local Chinese in Southeast Asia were often tense and bitter. Yet they were generally much smoother with other groups, especially Malays. Japanese assistance to local liberation movements in Indonesia and Burma later proved strategic in their efforts to throw off European colonial rule.

In wartime Indonesia, then the Dutch East Indies, Japanese imperial troops first trained the nationalist fighting force, two of whose key members were president-to-be Suharto and longtime foreign minister Adam Malik. They turned over critical stockpiles of weapons to the Indonesians during the hiatus between their surrender and the return of the Dutch. This move gave life to the protracted independence struggle that culminated in Dutch recognition of Indonesian sovereignty late in 1949. Post-independence ties have been reinforced by trade and aid—to the point that Japan has been Indonesia's largest trading partner and Indonesia has been the largest recipient in the world of Japanese aid—for most of the past four decades.

In Burma, Japanese intelligence was active even before the onset of war, training and equipping the so-called Thirty Comrades, the heart of the independence movement and Burma's postwar ruling elite. They were landed secretly by night on the Gulf of Martaban coast just as Japan's invasion was beginning. One of the Thirty Comrades, Ne Win, became the dominant Burmese leader of the postwar period, commanding the Burmese army and serving as president from

1962 through the 1970s and 1980s. Aung San Suu Kyi, the leader of Burma's opposition and winner of the 1991 Nobel Peace Prize, is the daughter of the man who first led the Thirty Comrades ashore. She herself has periodically traveled to Japan and once studied at Kyoto University's Institute of Southeast Asian Studies. Despite Burma's general policies of socialism and isolation, economic ties grew remarkably intimate under Ne Win. Indeed, Burma (now Myanmar) has frequently been among the five largest recipients of Japanese foreign aid.

Thailand's wartime experience with Japan was also broadly positive. Thailand formally remained neutral throughout the war, but cooperation with the Japanese military was close, and Imperial Army troops were stationed on its soil. Like Japan, Thailand has the distinction of being one of the few East Asian states never to fall under Western colonial rule. Together with shared traditions of Buddhism and monarchy, this long-standing independence has provided a common bond between Thailand and Japan that has transcended economic expediency. Not surprisingly, Thailand was, together with Indonesia and Malaysia, one of the sites chosen in 1991 for Emperor Akihito's first state visit to Southeast Asia. Such "imperial diplomacy" is a major and increasingly important element in Japan's efforts to strengthen ties throughout the region, symbolizing the end of the era represented by the previous Emperor Shōwa (known in the West as Hirohito), in whose name the invasions and occupations of World War II had been launched.

Vietnam is another highly strategic, non-Chinese-oriented nation of Southeast Asia where the experience of World War II with Japan has broadly cooperative implications for the future. As in Indonesia, defeated Japanese troops at war's end surrendered many of their arms to local nationalist movements rather than to returning European colonialists, thus fueling the independence struggle. Ho Chi Minh's active efforts to oppose the French return to Indochina in late 1945 were greatly aided by Japanese support. That is a reality his successors in Hanoi have never forgotten, which aids Japanese economic diplomacy even now. Indeed, many of the Japanese traders dealing most intimately with Vietnam today are either former Imperial Army soldiers posted there or their descendants.

In the postwar period, heavy Japanese investment in Southeast Asia has strongly supported that region's growth, especially since the mid-1980s. By 1994 the massive Japanese investment in ASEAN, for example, was well over $35 billion, or nearly triple its already substantial level of 1985. In Thailand alone such investment had increased sixfold since the mid-1980s, to over $6 billion.[4]

Such investment has been much more limited in Northeast Asia; indeed, the growth of Japanese investment in much of this region, including South Korea, has stagnated over the past decade. To be sure, some important and growing outposts of Japanese business still remain, like Dalian in Liaoning province, where several billion dollars of Japanese capital is deployed. Such beachheads give Japan important residual economic interests in the region that complement its obvious geopolitical concern for territories so unsettlingly close to its own home islands. Yet overall, Japan and its neighbors tend to be economic competitors rather than collaborators, reinforcing the historically and geostrategically derived tensions of the Northeast Asian Arc of Crisis.

The Special Complexities of Rebuilding the Northeast

The dynamics of investment in Northeast Asia, once again, are at once both more unstable and more unhealthy than farther south. Apart from a Manchurian and North Korean industrial base that is now aging, there is little positive infrastructural heritage of colonial rule (in this case Japanese) for future economic development. Still-unresolved Cold War obstacles and the cumulative fruits of economic stagnation under socialism compound the problem. Railways, airports, and shipping facilities are in poor repair, requiring much more infrastructural investment than is true in the south. Private firms remain more cautious about investing on purely market grounds, making the role of government all the more crucial. Any high-growth Hong Kongs or Guangdongs that might emerge in Northeast Asia would have to be strongly encouraged by the state.

Yet the political obstacles, both domestic and international, to aggressive state support for Northeast Asian economic development are legion. The socialist-heritage nations of the region, especially Russia and North Korea, but to a substantial degree even China, are dominated by interventionist yet fragile central governments with only limited understanding of market principles. Their ideologues and bureaucrats fear both decentralization and deregulation—the essence of regional integration in this part of the world—as a mortal threat to their power and purity. Nationalistic sensitivities compound their resistance.

The nonsocialist nations—major prospective source of infrastructure and capital infusions for any conceivable Northeast Asian economic takeoff—confront their own internal obstacles to deepened interdependence, made devastating by the inevitable reliance of sustained growth in Northeast Asia on government activism. Within Japan, the major obstacle is the continuing Russo-Japanese standoff over the disputed Northern Territories. This is coupled with a perception that Russia is too unstable and unpredictable, with resources insufficiently attractive relative to other possibilities that Japan entertains, to justify the risk. In South Korea, the North-South dispute within Korea itself, coupled with the troubled prospects of northern Asia compared to the booming south, inhibit aggressive investment and impede strong government activism to support it.

Local governments could make an unusually important contribution to Northeast Asian economic integration, compared to patterns in other regions. Some such bodies, such as both Niigata prefecture and Niigata city in Japan, are making vigorous efforts to promote Japan Sea regional development. Yet despite progress in personnel exchange and small-business support, they remain handicapped by their scale and scope of operations in solving the large and ultimately crucial questions of infrastructure, state regulatory policies, and overall political support for expanded regional trade and investment.

The United States and Western Europe, in sharp contrast to the situation in Southeast Asia, have no residual presence in Northeast Asia dating from colonial days, when the region was a Japanese preserve. And devoid of established stakes, Westerners as yet have no strong incentives to develop new ones. They can do much better, with less risk, in the booming, relatively stable economies farther south.

Perils of the North Pacific Strategic Equation

Southeast Asia has, with the parallel Russian and American withdrawals from Vietnam and the Philippines, largely become an area where security problems are conventional, and regional in character. Across Northeast Asia, by contrast, the dual specters of nuclear confrontation and nuclear proliferation loom much larger—indeed, arguably larger than in any other part of the post–Cold War world. They have global geostrategic dimensions—indeed, a succession of U.S. CIA directors and Secretaries of Defense have called developments in the region not only a challenge to Pacific Defense, but the most dangerous threat to peace in the entire world. And all this on the very doorstep of Japan, by far the most consequential nonnuclear power on earth, in economic, technological, and even military terms.

Ringing the Korean peninsula, linchpin of the Northeast Asian Arc of Crisis, three of the four major regional powers apart from Japan (China, Russia, and the United States) already forthrightly proclaim their possession of nuclear weapons. Only fifty miles from the North Korean border, around the sprawling naval port of Vladivostok, Russia has amassed a large local stockpile of tactical nuclear weapons, as noted in Chapter 1. Beneath the 600,000 square nautical miles of the Sea of Okhotsk, north of Hokkaido and west of Kamchatka, cruises 40 percent of the Russian strategic nuclear submarine fleet, with its ICBMs traditionally targeted, throughout the Cold War, on the major population and economic centers of the American Pacific coast.

Indeed, the Sea of Okhotsk still serves, together with the Barents Sea adjoining Finland and Norway, as one of the two major strategic sea bastions for Russian SLBMs. Its vast expanse and protected location, sheltered from deadly foreign killer submarines by the Siberian landmass, the Kuril Islands, and rugged Kamchatka, give that lonely sea an enduring strategic significance for Russia that transcends the Cold War. Its value should persist as long as Russia retains major strategic nuclear capabilities.

To protect the Sea of Okhotsk, Russia has ringed it with fortifications, missile defenses, and communication systems that remain substantial even as the Cold War wanes and some equipment begins to rust. With approximately sixty-five major surface combat ships and seventy submarines, fifty of them nuclear, the Pacific fleet based at Vladivostok is the largest of Russia's naval fleets, with responsibilities as far afield as the Arabian Sea.[5] It has assumed special importance as the second largest, the Black Sea fleet, has been immobilized by political conflict with Ukraine.

Aside from the Delta III–class SLBMs cruising the Sea of Okhotsk, Russia has both ICBMs and strategic bombers deployed along the Trans-Siberian Railway. In all, between one-quarter and one-third of Russia's strategic missile forces are deployed in the Far East. From its air base at Dolinsk Sokol, near the southern tip of Sakhalin, the Russian air force has strike capacity with MiG-31 Foxhound strategic interceptors against the whole of the Japanese archipelago, as far as the southernmost island of Kyushu, while Su-24 and Su-27 long-range aircraft based on Vladivostok have similar potential. The capability of Russian forces in the Far East has also been enhanced since 1991 by the addition of IL-76 AWACS planes.

Dangerous Korean Uncertainties

The situation on the Korean peninsula remains even more unsettling, despite some tentative recent steps toward peace. Behind the huge standing armies of North and South—in aggregate larger than the forces of either Cold War superpower—stand nearly three times as many men in reserve. Universal conscription has been standard in both halves of the peninsula since the Korean War; draftees serve a minimum of twenty-six months in the South, and a miserable five to eight years in the North, before reserve duty that extends in the North to the age of sixty.[6] North Korea spends over a quarter of its GNP on defense, the highest share in the world apart from Serbia, and five times that of the United States. South Korea, in response to the North's challenge, devotes a 3.8 percent portion of its much larger economy to defense. That is higher than any European member of NATO apart from Britain, Greece, and Turkey.[7]

Given North Korea's collapsing economy, grievously undermined by the disintegration of the Soviet Union in 1991 and its related diplomatic isolation, that reclusive nation claims virtually no leverage in international affairs apart from its ability to threaten others militarily. It has two clumsy, yet brutal and potent options: reckless forward deployments and a drive for nuclear weapons.

Today thousands of North Korean long-range artillery pieces, as noted earlier, peer starkly out from underground bunkers, southward toward Seoul. Most have highly automated firing mechanisms that allow them to emerge from massively fortified positions for only split seconds, denying South Korean and American artillery spotters time to launch a destructive counterstrike. The 600,000 North Korean troops within fifty miles of the DMZ, nearly 70 percent of the entire army, live amid a vast subterranean network of tunnels designed to neutralize American airpower, quietly awaiting the command to strike. At their side are deployed the vast bulk of the 10,000 SAM missiles, 4,200 tanks, and 2,500 personnel carriers at Pyongyang's command.

Attacking southward would obviously be suicidal. Yet the credible threat to do so is one of the few potent weapons in the North's limited diplomatic arsenal. Little, even the marginal relaxation of tensions following the October 1994 U.S.–North Korean nuclear agreement, can alter this grim logic of countervailing economic and military power, so starkly manifest on the DMZ.

A similar calculus of the cornered and the defiant also decrees a North Korean nuclear program that other neighboring nations can only regard as threatening.[8] Tactically, the intense concentration of forces along both sides of the DMZ increases the incentives for nuclear escalation in an extremity, especially should the defensive side have such weapons and the offensive side lack them. Pyongyang's own possession of nuclear weapons would increase the prospective cost of such a U.S.–South Korean escalation and thus help to preserve the overall credibility of the North's conventional threat. At the strategic level as well, nuclear weapons would provide invaluable leverage for beleaguered, economically weakened Pyongyang, especially if it could develop advanced delivery systems. North Korea also suffers from acute energy shortages, for which nuclear power, including plutonium-generating fast breeder reactors, provides a natural solution.

Classified assessments done by the CIA, the Defense Intelligence Agency, the National Security Agency, and the Energy Department all concurred by 1993 that North Korea most likely possessed crude nuclear weapons. It fashioned them, so the reasoning went, by reprocessing plutonium withdrawn from the Yongbyon nuclear reactor, sixty miles north of Pyongyang, during a hundred-day period that it was shut down during 1989.[9] The CIA 1993 National Intelligence Estimate suggested that up to twenty-six pounds of plutonium, or enough under optimum conditions to make two nuclear bombs, could have been extracted from the reactor during that time.[10] Other assessments have suggested that the North by 1994 possessed six nuclear devices or more.[11]

Drawing its own uranium concentrate from the Sunchon-Wolbingson mine near the Yellow Sea just north of Pyongyang, North Korea could potentially accumulate raw material for several nuclear weapons, in the absence of persistent foreign monitoring and political suasion. It has virtually completed reactors and reprocessing facilities that, were they to become operational, could produce enough nuclear material to build more than ten nuclear bombs a year.[12] American intelligence officials have also detected craters near the nuclear site at Yongbyon consistent with experiments to develop the high-explosive conventional munition triggers necessary to detonate a nuclear bomb. Clearly North Korea already has some significant residual nuclear capability, mostly new and developed since 1989.

In October 1994 the United States and North Korea concluded an agreement that would first freeze and then gradually transform the North Korean nuclear program away from its apparent military focus, over a period of five years. The agreement would forestall reprocessing of eight thousand rods of spent nuclear fuel from the Yongbyon reactor, containing five bombs' worth of plutonium, and ready it for eventual removal. It would also allow inspection of declared nuclear sites and freeze construction of two other dangerous graphite-based reactors in the North. These include one at Taechon, near the Chinese border, that was previously scheduled to begin producing ten times as much plutonium as Yongbyon in 1995. In return for North Korean concessions with respect to its nuclear program, the United States agreed to supply the North with badly needed heavy oil for its electrical generators over a period of several years.[13] It also agreed, in cooperation with South Korea and Japan, to build the North two $5 billion light-water nuclear reactors over the ensuing decade, to supplant more dangerous, weapons-capable facilities being frozen under the agreement.

The 1994 accord, to be sure, addresses the deep energy insecurities that are at least one major motive behind Northeast Asia's quest for nuclear power. The agreement does, in the short run at least, inhibit further military nuclear development that the world has no easy alternative means of suppressing. But its multi-year staged implementation clearly takes major calculated risks regarding North Korean compliance. Those risks are made especially difficult to quantify by the domestic political uncertainties within North Korea generated by the death of longtime president Kim Il Sung in mid-1994.

The agreement promises North Korea free oil, new reactors, aid, and the removal of many extensive existing trade sanctions. Most important, the prospect of substantial residual North Korean military nuclear capacity remains intact for at least several years into the future. Indeed, there is no clear provision for definitive inspection of the most suspicious nuclear sites until as late as 2003.

Independent of the 1994 agreement, major elements of Pyongyang's nuclear weapons program, such as designing a nuclear weapon or fabricating such weapons from materials already in stock, will likely continue.[14] Even if the North should scrupulously observe the nuclear agreement, it also retains other potent weapons of mass destruction. These could seriously threaten its neighbors, no matter what happens to its residual nuclear capacity. Since the late 1950s the North has, for example, developed an active and elaborate chemical weapons arsenal. By the 1990s this arsenal prominently included sarin, the deadly nerve gas developed in Nazi Germany that killed 12 and injured 5,500 people on the Tokyo subway in March 1995. North Korea also stocks phosgene, the murderous mustard-gas-family member that killed hundreds of thousands in World War I, as well as blood agents such as hydrogen cyanide.

For over twenty years, Pyongyang has also been actively developing bacteriological weapons. These now reportedly include at least ten different strains of bacteria, such as anthrax, cholera, bubonic plague, and yellow fever. They also include multiple pest-spawned diseases.[15]

The major challenge remaining for the North Koreans—one crucial to their capacity to credibly project nuclear, chemical or biological threats internationally—is building a missile capable of reaching potential adversaries, either close by in the region or farther afield. This capacity was unconstrained by the 1994 nuclear agreements. And there is a significant danger that rising economic intercourse between North Korea and the outside world, unless carefully monitored, actually increases it, by providing strategic, dual-use microelectronic or optoelectronic technology that fills important gaps in Pyongyang's current capabilities. The North Koreans reportedly already use Japanese technology for their missile guidance systems, and Japanese police have been working hard to prevent further diffusion.[16] The Iranians have also reportedly been providing major financial support.[17]

In May 1993, North Korea for the first time successfully launched three Rodong 1 surface-to-surface missiles, deep into the Sea of Japan. Rodong 1 is capable of carrying biological, chemical, or nuclear warheads, all of which the North Koreans are said to possess, and has a 650-mile range, enabling it to hit Osaka, Japan's second-largest city. On Pyongyang's drawing board is the two-stage Rodong 2 missile, designed with Tokyo clearly within its roughly thousand-mile range.

Over time, North Korea's weapons delivery capabilities may well continue to strengthen. Since 1990 its engineers have pursued joint ballistic missile development with the Iranians and the Chinese, at heavily guarded facilities near Iran's ancient capital of Isfahan. Through independent research, reverse engineering, and this sort of covert cooperation among pariah states, Pyongyang hopes to develop a new generation of two-stage intermediate-range missiles, the Taepo Dong 1 and 2. These missiles, based on fully proven Chinese M-11 technology, would be able quite accurately to hit targets all over Japan, as well as Beijing, Shanghai, and Vladivostok, if only problems with pumps for the clustered rocket engines and stage-separation technology can be mastered.[18] North Korea's ambiguous but rather comprehensive nuclear and CBW threat, in short, has become one that should give pause to Japan and other adjacent nonnuclear powers, thus heightening tensions and pressures for further arms proliferation around the Northeast Asian Arc of Crisis.

Nuclear uncertainties on the Korean peninsula stretch far beyond the prospect of a North Korean attack elsewhere in the region. They are also profoundly linked to, and amplified by, the clouded long-run political prospects for the regime in Pyongyang itself. The North Korean government may well suddenly collapse, as did its Communist counterpart in East Berlin during 1989–90, with Korean reunification following close behind, on the German model. Should this occur, Pyongyang's nuclear, chemical, and biological capabilities might suddenly fall into the hands of a South Korean regime with many of the complementary skills in missile propulsion and guidance systems, as well as telecommunications and precision electronics, that the North Koreans apparently still lack. Whatever its actual intent, a united Korea could thus pose a different, and deeper, prospective nuclear challenge to its neighbors than even the bitterly divided and heavily armed Korea of the present.

Sudden collapse of the North Korean regime could entail destabilizing economic as well as political costs. A virtually inevitable reunification under such circumstances would cost South Korea as much as $1.2 trillion, according to that nation's 21st Century Committee.[19] Sudden unification would result in unemployment of as high as 30 percent in the North and an exodus of 4.5 million North Koreans to Seoul, effectively immobilizing that capital city of twelve million people.

Apart from the domestic social turbulence, reunification could also trigger serious transnational tensions with Japan and its volatile Korean minority of 700,000, as well as China, whose border areas are home to over a million Korean Chinese. Lightly armed Japan could easily feel threatened by the merging of two muscular Koreas that together boast a total of nearly two million men under arms, especially if one or both held residual nuclear capabilities. The huge financial demands of North Korean reconstruction would naturally attract Japanese capital.

Yet the terms of lending, as well as conditions of access for Japanese business, could be bitterly contested issues, given Japan's long history of exploiting Korea. In a newly unified Korea, American bases might well seem superfluous. But an American withdrawal could intensify ancient Sino-Japanese rivalries over Korea, simultaneously weakening the legitimacy of U.S. bases in Japan as well.

A different danger is posed by a more ambiguous and stormy passing of Kim Jong Il's regime: the specter of civil war. Should rival factions clash violently for supremacy, in a confused power struggle to succeed Kim, the command and control system for North Korea's nuclear capacity might well break down. One or another faction might, in such a struggle, resort to the grimly decisive power of nuclear or CB weapons. Such possibilities do distinctly loom in the North Korea of the future, although their probability is difficult to determine.

Other Flash Points of the Arc of Crisis

The prospects for Korea's political future are troubling under virtually any scenario, despite its strong economic record and future potential. Across the at once promising and forbidding lands of Northeast Asia, however, Korea is hardly alone as a tinderbox. At the south of the arc lies the Taiwan Strait, and the prospect of recurring tensions between China and Taiwan. There has been a substantial recent buildup on both sides of the strait, even as economic intercourse has deepened. The 1992 election-year Bush administration sale of 150 F-16s to Taiwan, followed by France's parallel sale of sixty Mirage 2000-5s, accelerated the buildup. China's steady development of stronger blue-water naval capabilities, as its reliance on imported Middle Eastern oil rises, gives the buildup added complexity and momentum.

THE TAIWAN STRAIT

Many military flash points in the Pacific can be stabilized by international agreement, or by the emergence of a new regional security framework. One of the dangerous aspects of the China-Taiwan confrontation, so starkly manifest in the volatile cycle of missile rattling following Taiwanese president Lee Teng-hui's U.S. visit in mid-1995, is that it cannot. Even more than the intra-Korean confrontation—a civil war in extended, but precarious, suspension, which at least involves the stabilizing presence of the United States and the United Nations—the Taiwan Strait conflict cannot easily brook outside mediation. As China grows, both economically and technologically, Taiwan must either keep up its military guard or risk being savaged, like a weak younger brother in a brutal family quarrel for which no enforceable rules exist.

There is, to be sure, a dual cooperation and conflict character to the Taiwan Strait confrontation—China's ongoing civil war—that observers must not lose sight of. The main protagonists, after all, are all Chinese—many of them in deepening communication with one another. Since mid-1993, even ordinary letters have been flowing back and forth across the strait, amid all the rhetoric. The two sides make well over thirty million phone calls a year to one another. Cross-investment exceeds $20 billion, and bilateral trade is well over $15 billion annually.

Yet this very interdependence raises both the stakes and the prospective consequences of conflict, both for Chinese themselves and for vitally interested third parties, especially Japan, which lies next door. The simple existence of interdependence is no guarantee that serious conflict will not occur. The American Civil War demonstrated that reality just over 130 years ago.

Deng Xiaoping once declared that five conditions would provoke a Chinese military attack on Taiwan: (1) a Taiwanese nuclear capability; (2) a Taiwan-Russia entente; (3) an outbreak of extreme political disorder in Taiwan; (4) a declaration of Taiwanese independence; and (5) a rejection of unification talks "for a long time."[20] The most credible of these threats may well be that to retaliate against a Taiwanese nuclear breakthrough. And Taiwan is clearly, as noted in Chapter 4, close to the nuclear threshold. Although the dangers of a Taiwanese-Russian entente are now limited for China, given the collapse of the Soviet Union and China's own deepening ties with Moscow, political transition in Taipei could also be the catalyst for serious escalation of tensions across the Taiwan Strait.

A flowering of the Taiwanese independence movement, in particular, would present the Chinese leadership in Beijing with the prospect of a Taiwan decisively torn from the motherland—an eventuality that they have, with a profound sense of destiny and history, categorically refused to accept. In the Chinese way of thinking, acceptance of such conclusive separation would disgrace the name of those permitting it throughout history, from generation to generation. Some, however, reject the need for such culturally based explanations, asking bluntly: Did not Abraham Lincoln himself reject secession in America's own Civil War?[21]

The prospect of violence across the Taiwan Strait would be especially serious in the context of severe centrifugal pressures within China. These appear likely to emerge, as its explosive current growth leads to greater regional autonomy and deeper income gaps between have and have-not areas. Generational transition in Beijing's leadership is also adding, of course, more uncertainties.

If the unity of China as a whole is to be maintained, Taiwan cannot be allowed to transparently go its own way. Yet there is real prospect of the Taiwanese independence movement gaining greater momentum, more directly defying the ruling Kuomintang authorities, and in the process gravely vexing Beijing. The independence-oriented Democratic People's Party (DPP) won the mayoralty of Taipei, Taiwan's capital and largest city, in December 1994 local elections. They will be a major factor in future direct elections for the president of the Republic of China. Should an avowed advocate of independence at some point actually win, that could provoke a serious crisis with China.

Even without explicit national-level victories by the DPP, Taiwan's democratization—and continuing desire for autonomy from the mainland—could exacerbate tensions across the Taiwan Strait. The DPP's appeals themselves have encouraged top Kuomintang leaders, including Taiwan-born president Lee Teng-hui himself, to make increasingly forthright and activist efforts to secure international recognition for Taiwan, in an effort to compete politically within Taiwan itself. And Taiwan's democratization, contrasting so sharply to continuing repression on the mainland, also gives Taiwanese leadership the confidence and legitimacy to speak out.

These dynamics culminated in President Lee's May 1995 visit to the United States, and Beijing's vehement reaction, including four extended tests of nuclear-capable M-9 IRBM missiles in an ocean target zone only eighty-five miles from Taiwan. The Taipei stock market dropped 3 percent in a single day, and 30 percent in a matter of weeks, as Taiwan responded with its own military exercises, and President Lee pointedly suggested the possible need for a review of Taiwan's dormant military nuclear program. The 1995–96 presidential campaign on Taiwan further worsened China-Taiwan relations by encouraging a Taiwanese assertiveness in international affairs especially provocative to a Beijing itself in transition.

What could happen if Taipei, only twenty minutes by F-16 southwest of U.S. bases in Okinawa, were to seriously upset Beijing? The tactical situation is just complex yet just balanced enough to make for tragic miscalculation. Taiwan's geographic location—ninety miles of open water stretching between it and the mainland, coupled with a rugged, mountainous topography—strongly favors its defenders. Indeed, during World War II the United States estimated that it would take 300,000 American troops to defeat 32,000 Japanese ground troops then occupying Taiwan—nearly a ten-to-one ratio.[22]

Taiwan also has a powerful military of 270,000 soldiers and 30,000 marines, as well as extensive reserves, to buttress its defense. The air force is going to extraordinary lengths to reinforce itself, as indicated both in its huge F-16 and advanced Mirage acquisition program and in its extraordinary Chiashan project. On the remote southeastern side of Taiwan, farthest from the threat of Chinese planes and missiles, Taiwan is building an elaborate underground system of tunnels and interconnected aircraft hangars, fuel lines, and storage depots to ensure that Taiwan's fighters have survivable bases. At least two hundred of its advanced fighters will be housed at Chiashan when this massive redoubt is completed around the year 2000.[23]

Yet mainland China confronts Taiwan with nuclear weapons, three million men under arms, ICBMs, more than sixty times Taiwan's population, an economy that may well be the largest in the world within two decades, and no small supply of hubris. At some point could it not miscalculate? What would it be most tempted to do?

Most military analysts see a naval blockade as being the most likely form of serious Chinese military pressure on Taiwan. Taiwan, after all, is a relatively small island that is heavily dependent on international trade. In 1992, for example, its combined exports and imports came to 78 percent of GNP, or more than $153 billion, just 8 percent less than the entire direct trade of the Chinese mainland.[24]

Taiwan is, to be sure, largely self-sufficient in food and holds at least two months' reserves of critical resources like petroleum. But even a declared or lightly enforced blockade could gravely damage its economy by raising insurance rates and making many ships unwilling to call at the island's newly vulnerable ports. Beijing could rather easily deploy submarines to demonstrate its determination to enforce the blockade, with potentially disastrous consequences for Taiwan's foreign trade lifelines.

Precisely because of its vulnerability to a high-intensity blockade, Taiwan would be forced to respond to such a Chinese move with an immediate escalation of the conflict. This could include using Taiwan's air force against mainland shipping and ports. It could also involve mining major ports such as Shanghai, or attacking Fujian province with missiles deployed on Nationalist-held islands in the Taiwan Strait.

China, in the final analysis, has an overwhelming superiority over Taiwan in men and material that can only grow with time. As China becomes an economic and possibly military superpower, this balance could well shift into the gray area where an attack or overt military pressure might seem rational to some. As with Israel in the Middle East, faced with large forces arrayed against it, Taiwan could develop incentives for a preemptive strike. Like the Gulf of Aqaba and Suez Canal of the 1960s and 1970s, the Taiwan Strait of the late 1990s could well become the hair trigger of an entire volatile region.

BEIJING AND BEYOND

From the Taiwan Strait the Arc of Crisis passes through the coastal cities of China—Xiamen, Shanghai, Qingdao, and Tianjin—to the very gates of central government power in Beijing. Now booming, these cities are attracting millions of immigrants from the countryside in search of a better life, with a huge number of potential rural migrants behind them. Many professional estimates suggest that as agricultural productivity improves in the countryside, where 80 percent of Chinese now live, as many as 200 million underemployed farmers could flood to the cities, mostly the coastal cities of the Arc of Crisis.

Moving north and northeast across the arc from Beijing, which has troubles of its own better discussed elsewhere, one approaches the windswept frontier lands of Manchuria. For a generation from 1960, when Mao Zedong expelled Soviet advisers, these areas were deeply estranged from the Russians to the north. Even today one can visit the elaborate bomb shelters of Dalian, in southern Manchuria, complete with field hospitals, command posts, and even barbershops, which Red Guards prepared during the late 1960s in full anticipation of a Soviet nuclear strike. The fears were, as new historical evidence suggests, not unjustified; the Chinese and the Russians actually clashed on the Ussuri River in northern Manchuria several times during 1968–69, and several memoirs from the period suggest that Leonid Brezhnev indeed sought Richard Nixon's assent to a Soviet strike against Chinese military and industrial sites not long after Nixon took office.[25]

Manchuria, in contrast to booming southern and central coastal China, has a mixed economic present and prospective future. To be sure, there are pockets of buoyant prosperity, such as Dalian, the chief port of southern Manchuria. It began booming in the early 1990s, on the strength first of heavy Japanese foreign-aid investment in import and transshipment facilities, followed by a multibillion-dollar surge of Japanese manufacturing investment from late 1992.

Yet Dalian's prosperity, based heavily on local assembly and processing of imported inputs such as micromotor and camera components, has not spilled over to surrounding areas. Northern Manchuria, adjacent to Russia and North Korea, remains much more stagnant and troubled. Indeed, Jilin province, at the doorstep of North Korea, has a per capita income less than two-thirds that of neighboring Liaoning, where Dalian is located, and only one-third that of Beijing, just four hundred miles to the south.

Manchuria, until the Chinese Revolution of 1911 a sacred and underpopulated homeland of the imperial family whose economic development was considered sacrilege, became a heartland of heavy industry under the Japanese occupation and a base upon which Mao's fledgling government built further after 1949, with Soviet assistance. The steel of Anshan, the oil of Taching, the motor vehicles of Changchun, and the jet fighters of Shenyang became the very economic and military sinews of the Revolution itself across the 1950s and into the 1960s. But the same traits that had made Manchuria indispensable in the Korean War era have become an albatross. The Chinese economy has shifted toward light industry and electronics, with the Four Modernizations and the economic liberalization since Mao's death in 1976. Manchuria has not kept up.

Manchuria, especially the two northern provinces of Heilongjiang and Jilin that abut Russia and North Korea, has three deep, intractable structural adjustment problems. Much of its industrial base is old and outmoded, dating from the 1950s and before. Changchun's auto plant, for example, was built in 1956, and its chemical complex, the largest producer of synthetic ammonia in China, began operation in 1957.[26] Many of the basic industries of the region, such as coal and steel, have been superseded by newer sectors such as electronics, which have flourished elsewhere. Others, such as Heilongjiang's venerable Taching oil fields, which still produce one-third of China's crude oil, face steady depletion.

To make matters worse, most production in Manchuria is concentrated in grossly overstaffed public corporations and state farms, increasingly forced to lay off workers as the region's economy gradually and painfully liberalizes. Indeed, probably one-third of all jobs in such Manchurian enterprises are currently surplus.[27] Furthermore, these public corporations, as in the former Soviet Union, typically dominate their towns and sustain the broader community, whose hospitals, schools, and day care centers will face grave cutbacks as the cold claims of rationalization proceed.

Beyond the structural adjustment problems that a heavy concentration of public corporations and outmoded heavy industry present, Manchuria confronts a host of infrastructural and administrative obstacles that inhibit coherent regional development and integration with the broader North Pacific region. For one thing, local transportation is extremely poor; the interior of Manchuria still communicates with the world largely over a single railway line from north to south, built by the czarist Russians at the beginning of this century. Roads and feeder railways are both decrepit, while impoverished Jilin and Heilongjiang provinces lack their own outlets to the sea.

As in so many socialist nations, red tape, poor cooperation among localities, and lack of clear market incentives make it excruciatingly difficult to correct the distorted, often irrational patterns of regional specialization that now exist. All economic planning in China is at either the national or the provincial level; there is no systematic regional development planning whatsoever, in Manchuria or anywhere else. The three provinces of Manchuria have tried to develop a common master plan for the year 2000, but failed due to political jealousies. To make matters worse, there is no unified market to create economic pressures for rationalization, a common malady across all of continental Northeast Asia.

The result is interminable red tape, making it difficult for firms and labor to move to where the market needs them. This situation also breeds duplication of investment, chronic inflationary pressures, and openings for increasingly powerful local mafia and overseas Chinese groups. They often seem to be the only actors capable of correcting the static rigidities that are a major contribution of China's northeast to the Arc of Crisis.

Mounting economic stagnation and structural difficulties in the smokestack industries dominating the Chinese northeast have begun to provoke unsettling difficulties for the broader Northeast Asian region as a whole. Most important, they have intensified incentives for individual Chinese of the region to press outward and to seek their fortunes across China's porous Siberian land frontier. Several hundred thousand illegal Chinese immigrants are reportedly squatting in the sparsely populated Russian Far East, in whose three-million-square-mile territory—the size of the United States—only eight million Russians live.[28] Virtually all of these intruders have appeared since the collapse of the Soviet Union, and the dissolution of systematic border formalities, from the beginning of 1992. During 1994–95, local Russian authorities made strenuous efforts to control and ultimately to evict them, leading to an escalation of tensions along the frontier.

Russia and China: The Uncertain Jousting of Giants

It is in the Russo-Chinese relationship—at both the local and the national levels—that the turbulent and momentous dynamics of the Northeast Asian Arc of Crisis are at their most complex and fateful. The border of these two giant nations twists four thousand miles across the heart of Asia, from the jagged Altai mountains of Sinkiang to the stormy North Pacific. In the post–Cold War world, with both nations in deep ferment, these borderlands stand amid historic and fateful transformation.

Arms and technology flows from Russia to China are a major element of the change now under way. These flows, including MiG-31 strategic interceptors, Su-27 fighters, Tu-22 bombers, T-72M main battle tanks, and A-50 airborne warning and control planes, are, to put it mildly, not strictly necessary for defensive purposes.[29] These flows also provide much-needed foreign exchange for Russia and encourage a high priority on ties with China in the Kremlin.

Relations at the local level have become much more tense, however, especially in the Russian Far East (RFE), even as economic interdependence has risen. In the wake of Boris Yeltsin's January 1993 visit to Beijing, Chinese investment began pouring into the Far East, especially the Vladivostok area. During that month, investment by Chinese companies in Vladivostok alone constituted 61 percent of all the foreign direct investment in Russia.[30] Local trade relations between northern China and the RFE have also rapidly intensified, as the RFE has been cut off from its traditional sources of supply in European Russia by the confusion in economic relations among Russian regions, as well as by spiraling, ever more market-oriented transportation costs across the huge distance of Siberia.[31]

By the end of 1992, the overall share of international trade in the RFE regional economy was double its level only two years earlier. Nearly one-third of that trade was with China alone. This Chinese share had reached three times its 1985 level and included a full half of the RFE's imports.[32]

Rising Russian resentment of the Chinese in the Far East has been fueled by what many Russians perceive as the exploitive nature of the mutual relationship. Most Chinese, responding to the acute shortages of consumer goods in the Far East following the Soviet collapse, have been preoccupied with short-term trading, often of low-quality goods that are sold at what are perceived as extraordinary prices. In 1993, according to unofficial data, almost $50 million in hard currency was transferred from Russia to China, largely by such traders.[33] Another significant proportion of the proceeds is reportedly being used to buy up local Russian real estate. But given the confused state of local police powers and statistical record keeping, no one knows for sure. Fear of Chinese smuggling and other criminal activity has also become rampant, leading to the widespread crackdowns on the Chinese during 1994–95.

The troubled response of the Russian Far East to economic interdependence with China since 1991 points to the larger Russian sense of vulnerability and distinctiveness in relation to Asian neighbors that has persisted since the Mongol invasions seven centuries ago. This Russian sense of unease, often overtly racial, inflames and complicates relations not only with China but also to an important degree with Korea and Japan. It has six major dimensions that, when taken in conjunction, suggest that the Russian sense of vulnerability may well deepen in coming years, with important destabilizing implications for both economic integration and political relations across the Northeast Asian Arc of Crisis as a whole.

The RFE feels itself vulnerable, first of all, because of its extraordinarily small population of Russians, virtually swallowed up in the vast expanse of Siberia. The Far East as a whole, as pointed out earlier, has only eight million Russian inhabitants, as compared to 1.2 billion in neighboring China, and over sixty million in

Manchuria alone. The RFE, furthermore, is separated from the rest of Russia by vast Siberian distances. Since 1991 those expanses have suddenly grown much greater with the reduction of government transportation subsidies and a consequent spiraling of air fares, freight costs, and telecommunications rates. Moscow is ten hours' flying time—seven time zones, as compared to only three across the entire continental United States—from Vladivostok. Russians on the Pacific coast naturally feel isolated and alone as they find it harder and harder economically to remain in touch with the rest of their own country.

Unresolved territorial disputes compound the perceived vulnerability of the RFE, and Moscow's defensiveness on its behalf. Despite the ongoing efforts of a border commission that has been working for more than a decade, Russia and China have come to no consensus on where their common frontier runs. Current territorial divisions are based on the 1860 Treaty of Beijing—one of the "unequal treaties" of imperialist days. China continues to regard this czarist relic as coercive and invalid. China's pique is intensified by the increasingly significant economic consequences. These include the enforced separation of depressed northern Manchuria from the sea at Zarubino, and from lucrative export markets in the rest of booming East Asia that such arguably unjust territorial separation implies.

The problem of territorial ambiguity is compounded for Russia by the enormous shift in relative economic fortunes that has been under way between China and Russia since the late 1980s. During 1981–87, just as perestroika was beginning in the Soviet Union, for example, Chinese and Soviet growth rates (4.5 and 1.7 percent respectively) were roughly comparable. Yet they have since diverged dramatically. For the 1985–91 period, Chinese growth stood at 12.7 percent annually, while that of the Soviet Union actually *declined* by a total of 18.2 percent over the same period.[34] During 1992–93 alone, the Chinese economy grew by roughly one-quarter, while the post-Soviet Russian economy contracted by about the same amount. China's GNP is now closing the gap with Russia at a speed neither Russians nor Chinese dreamed possible a decade ago.

The RFE's gnawing sense of relative economic vulnerability is also aggravated by the enormous structural adjustment difficulties that it still confronts; these contrast sharply, in Russian eyes, with China's more favorable situation. Most important, 40 percent of the RFE workforce in 1993 continued to be employed in military industry, although this represented only 12 percent of industrial production. Military firms were thus keeping huge numbers of redundant workers on their payrolls, even as they cut production—a staggering burden that clearly could not be sustained. They would have either to lay off workers, diversify toward new sectors, or find new markets for expanded military production.

As defense conversion was not progressing rapidly, the options increasingly were the two most destabilizing: layoffs or expanded military production, primarily for export across the Ussuri into Manchuria, which would ultimately only intensify Russian vulnerabilities to Asian neighbors. By the mid-1990s, this arms export option seemed to be the line of least resistance and greatest profit that the region was pursuing, despite its potentially counterproductive longer-term consequences. This was evidenced by Russian transfers of Pacific fleet submarines to North Korea, diverse arms components to China, and even two old Kiev-class aircraft carriers to South Korea.[35]

The Russian Far East is internally divided vis-à-vis the outside world, which also deepens local insecurity. Several of the major urban centers of the region—Vladivostok, Khabarovsk, and Yuzhno-Sakhalinsk, for example—struggle intensely with others for scarce central government resources and, increasingly, what inbound foreign investment does exist. This situation allows outsiders, including Chinese merchants across the Ussuri in Manchuria, to play local jurisdictions against one another, heightening levels of local Russian resentment, without provoking an effective counterreaction.

On top of all the psychological and economic anxieties that the RFE confronts, it is also strategically vulnerable—a consideration that plays heavily in Moscow as well as in the Far East itself. Virtually all of the region's population is concentrated in a small number of urban centers that could be easily and grievously damaged by hostile strategic air strikes. Both those population centers and the major transportation lines run cheek by jowl with the Chinese border. The Trans-Siberian Railway, for example, provides the only lateral land route along Russia's many-thousand-mile frontier with China and terminates at its major naval base in Vladivostok. Yet the railway runs at some points less than two miles from the Chinese border on the Ussuri River, through flatland with no natural defense points other than the river itself.[36]

The Russian Far East is simultaneously vulnerable and extremely important to Russia, a combination that naturally stirs a strong protective response from the motherland more than three thousand miles away across the Urals. The RFE is Russia's outpost on the Pacific, the world's most rapidly growing region; indeed, Vladivostok, Russia's preeminent Pacific port and terminus to the Trans-Siberian Railway, is literally "Commander of the East" in the Russian language. The RFE also has considerable geostrategic value, in that the Sea of Okhotsk remains one of the principal Russian nuclear sea bastions—critical to the invulnerability of the Russian nuclear deterrent. Finally, the region has enduring symbolic importance, emblematic of Russia's past imperial glory. It evokes a nationalist response to which Kremlin leaders, especially in the wake of challenge from the right in the late 1993 parliamentary elections and the muddled 1994–95 Chechnya intervention, have become eminently vulnerable.

Given the simultaneous importance and vulnerability of the Russian Far East—conditions that are both likely to persist or intensify into the future—Russia will have difficulty resolving its twin territorial disputes with China and Japan through concession or compromise. This uncertainty will leave a bitter residue of lingering tension, especially between Russia and Japan, where a peace treaty formally ending World War II remains unsigned more than fifty years after the guns were silenced. The two rising East Asian superpowers, for their part, are growing ever more self-confident, and simultaneously disinclined to acquiesce in seemingly inequitable relations with Russia.

Rising nationalism in the RFE, especially backlash against the increasing economic presence of China in the region, may also reinforce this Russian rigidity and raise the overall level of regional unease in Northeast Asia. Slow economic growth and inadequate incentives will inhibit defense conversion, creating the strong possibility of the RFE's becoming an increasingly important source of military exports, especially of defense components to China and the volatile North Korean socialist hermit kingdom next door. Russia's Far East, in short, will likely fuel still further the already volatile cauldron of tensions that is the Northeast Asian Arc of Crisis.

The Troubled Seas and Northeast Asia

Apart from its own instabilities and intraregional tensions, Northeast Asia also relates to Asia as a whole in ways that are more broadly destabilizing. Its major nations—Japan, China, and Korea—trade over vulnerable, extended sea trade routes that meander thousands of miles south and west around Southeast Asia and through the Indian Ocean, toward vital markets and oil supplies in Europe and the Middle East. Those sea routes traverse narrow straits that border Indonesia, Malaysia, the Philippines, and Singapore, as well as island possessions of India and Myanmar, before reaching the Indian Ocean.

Those sea-lanes are highly vulnerable to disruption at any one of many geographical choke points, as Figure 2-2 makes clear. And the major Northeast Asian powers have incentives to ensure that neither a local Southeast Asian power nor one of their number in Northeast Asia does so. With China beginning to develop a blue-water navy and South Korea taking cautionary steps, the long-term danger is rising of a classical naval arms race among Northeast Asian powers, particularly China and Japan, especially should the U.S. presence in the western Pacific decline.

Figure 2-2

THE VULNERABLE SEA-LANES OF EAST ASIA

Conflicting offshore territorial claims, all the way from the Sea of Japan to the waters just north of Indonesia, create another element of broader regional insecurity. Again, the economically and militarily powerful states of Northeast Asia are the principal catalyst for a broader, regionally destabilizing pattern of competition. They have extensive, competing offshore claims, the most important of which is the Senkaku Islands dispute between Japan and China in the East China Sea, southwest of Okinawa. China also claims huge portions (roughly 80 percent) of the South China Sea—areas that cut across Japan's sea-lanes, as well as the territorial claims of five other nations in Southeast Asia.

Why do these conflicting claims matter? Why are Northeast Asia's sea-lanes, and the defense thereof, so vital? Why is an Asia that in recent years has been so peaceful and prosperous now in danger of taking another, more troubling course? The answer lies heavily in a vital subject to which we now turn: *energy.*

LOOMING ENERGY INSECURITIES

There is a dark side to the explosive economic growth of East Asia, rooted in the region's profound energy insecurities. The United States has its Texas oil fields and its Alaskan North Slope. Western Europe has the North Sea, with its abundance of submarine oil and gas. Nearby Russia has some of the largest energy reserves in the world, as do Mexico and Canada, now linked through the new NAFTA accord with the United States.

Yet Asia, particularly troubled Northeast Asia, remains paradoxically destitute in energy terms, relative to the voracious demands of its highly geared industrial economies. Even including currently large or growing oil exporters such as Indonesia and Vietnam, the Asia-Pacific region provides little more than a tenth of global oil production, and less than a twentieth of world reserves. With a reserves-to-production ratio of only eighteen years, compared to the world average of forty-six and the Middle East average of 104 years,[1] East Asia also stands on precarious ground as it looks warily to the future.

Best-known, and most longstanding, are the energy vulnerabilities of Japan, which has never had even a single major developed oil field. This has not been for lack of trying. The first American geologists arrived in Japan, prospecting for oil, in 1869. During 1891 the small Amase field, near Niigata on the Japan Sea coast, began production, and several international majors prospected for oil before World War I. Yet they ultimately abandoned this high-cost venture; neither they nor domestic suppliers could meet more than a minimal share of the spiraling needs of the Japanese economy from local sources. In the mid-1990s, Japan still imported 95 percent of its oil.[2]

Coal mining commenced more than two centuries ago. Production rose steadily as Japan industrialized, to the point that Japan in 1936 became the third-largest coal-producing nation on earth.[3] But domestic coal supplies too were insufficient to meet the voracious demands of one of the world's highest-growth economies. As in oil, Japan has imported substantial amounts of coal ever since the early days of its opening to the outside world in the 1850s.

By 1980, Japan was dependent on foreign sources for a full 88 percent of its primary energy supply. Through vigorous conservation efforts and attempts to develop every sort of alternate energy—from solar power to thermal energy and the harnessing of the tides—it was able to cut its net reliance on imports below 85 percent by 1992.[4] But this remained higher than that of any other major industrialized nation on earth. Even with its gas-guzzling limousines and the steady depletion of onshore Texas oil fields, the United States imported only 18 percent of its energy. Britain, with its North Sea oil and gas fields, kept net imports to 3 percent. Germany, thanks to extensive use of domestic coal, imported 55 percent, and France, with nuclear power, held imports to 54 percent. No country but Italy was close to Japan's painfully high dependence on the outside world.

This energy Achilles' heel translated, by 1993, into a mammoth $53 billion Japanese fuel import bill, even with all the conservation of the 1970s and 1980s taken into account. Oil made up over half of Japan's total fuel imports, or twice Japan's steel exports, even at $20 a barrel. Were "black gold" to return to the stratospheric price levels of the 1970s, the burden on Japan would be far greater still.

Japan's energy vulnerability is painfully apparent whenever global oil prices rise. In 1973, after OPEC tripled the price of oil, for example, the usurious "Japan rate"—lending rates to Japanese firms at 1 to 2 percent above LIBOR, the London interbank rate—was immediately born, and the yen flew into a tailspin. The market response in recent years has been milder, but when Saddam Hussein's forces marched into Kuwait in 1990 the yen nosedived once again.

From an East Asian regional economic viewpoint, Japan's lack of energy resources matters greatly, because of the size of the Japanese economy and the voracious demands it thus imposes on international markets. As is clear from Table 3-1, Japan imports nearly three times more oil than any other nation in East Asia, and almost two-thirds of the liquefied natural gas moving in international

Table 3-1

ASIAN OIL HAVES AND HAVE-NOTS

A. *Oil Haves*

Country	Net Oil Import Dependency (%)	Oil Production/ Consumption (net balance, 1,000 bbl/day)	R/P Ratio	Oil Imports ($ billion/yr.)
Indonesia	−62.3	+840	12.2	2.1
Malaysia	−105.6	+415	12.5	1.5
China	−4.6	+405	22.6	2.1
Brunei	—	+165	33.2	—
Vietnam	8.5	+72	19.0	0.5

B. *Oil Have-nots*

Japan	95.4	−5,282	11.6	30.1
South Korea	63.3	−1,185	NA	13.0
Taiwan	44.6	−570	NA	3.9
Thailand	58.1	−440	NA	3.3
Singapore	100	−380	11	9.3
Philippines	75.8	−225	NA	2.0
Hong Kong	100	NA	NA	1.3

Sources. Economist Intelligence Unit; Far Eastern Economic Review, April Formenn, 1994 [4], [l]. [l]= 17; Asian Development Bank, *Key Indicators of Developing Asian and Pacific Countries*, 1994 ed., pp. 25–26.

Notes: R/P ratio equals years of reserves remaining at current rates of production. Figures for 1992, except Vietnam (1991).

trade as well. Since Japan makes up over half of East Asian regional GNP, even moderate Japanese growth does much to tighten regional markets. Since 1987, Japanese oil demand, for example, has risen 5 percent a year, adding 900,000 barrels per day to the region's requirements.[5]

Japan's endemic energy vulnerability has, of course, painful geostrategic implications that have often tormented Japanese policymakers in the past, and that had fateful implications for America's own Pacific Defense half a century ago. British and American officials, in response to Japanese aggression in Manchuria, did actually discuss imposing an oil embargo on Japan as early as 1934.[6] The first actual controls on oil supply to Japan, imposed in July 1940 upon the Imperial Army's invasion of Indochina, simply required export licenses for aviation fuel and scrap metals and did not explicitly name Japan. But they were construed in Tokyo as an effort to throttle Japanese economic and military strength amid a war with China in which the United States was nominally neutral, thus accelerating the rapid downward spiral in U.S.–Japan relations during 1940–41. The key role of energy as the lifeblood of Japan was affirmed dramatically in the Pearl Harbor attack, intended to cripple American defense of far-flung Pacific sea-lanes, and thus to open the way for Japanese advances toward the oil fields of Southeast Asia. A lightning paratroop attack on the Palembang oil fields in the Dutch East Indies followed almost immediately upon the December 7 attack on Battleship Row.

Other parts of Northeast Asia are even more precariously vulnerable in matters of energy than Japan. And their shortages, in many cases, are worsening rapidly. Most exposed, perhaps, is the Korean peninsula, whose economy is even more energy-intensive than that of Japan, and which has even less domestic energy at hand. South Korea, with large, energy-devouring petrochemical, steel, and shipbuilding industries, as well as a growing middle class ever more addicted to driving, has recently grown at 8 percent a year. But oil demand has risen 20 percent annually, and gasoline demand, propelled by a 22 percent annual increase in vehicle registrations since the late 1980s, has soared by 29 percent a year.[7]

Overall, South Korea tripled its per capita consumption of energy from 1975 to 1991.[8] This translated into an import bill of nearly $13 billion, or roughly triple the energy import bill, relative to overall national GNP, that Japan currently confronts. As a smaller nation, with less global political-economic clout than the economic giant across the Japan Sea, South Korea has had even more trouble competing for oil in times of shortage. This was graphically clear both during the two oil shocks of the 1970s and during the Gulf crisis of 1990–91.

The prospects are strong that South Korea's fears regarding energy security will deepen over the coming two decades. The Energy Institute of Korea forecasts, quite conservatively, that aggregate energy demand in South Korea will rise by 128 percent from 1992 levels by the year 2010, to 249 million metric tons of oil equivalent (Mtoe).[9] Even with conservation, an expanded nuclear program, and extensive use of natural gas, this would involve a sharp increase in South Korean oil imports, to 125 million Mtoe, by the year 2010. At that level they would come to more than half the imports of Japan, and double those of Taiwan and Hong Kong combined.

North Korea has substantial production and reserves of coal—over four times Japan's production, and nearly three times South Korea's.[10] Yet it has no oil what-soever, being traditionally forced to import the whole of its requirements either overland from China (75 percent of the total) or by uncertain and vulnerable sea routes more than seven thousand miles from Iran. If South Korea has problems obtaining scarce energy in times of shortage, these are magnified manyfold in the case of the North; it not only lacks foreign-exchange and geopolitical leverage, but also is excoriated as an international pariah. Even should it gradually deepen its international economic ties, any gains in energy supply would probably be offset by the rising demand that accelerated economic growth would generate. Korean reunification, of course, could greatly intensify all of Korea's energy prob-lems, especially if reunification led to a major, unforeseen acceleration of eco-nomic growth.

Chronic energy shortages, of course, have made the North's uranium reserves at Unggi, Pyongsan, and Hungnam, together with its controversial reactor and reprocessing plant at Yongbyon, even more important in the Korean peninsula's energy equation than they otherwise would be. They clearly intensify the North's attraction to nuclear power. The October 1994 American agreement to supply heavy fuel oil to the North was thus an important element in securing Pyongyang's assent to constraints on its nuclear program.

Taiwan also has major energy vulnerabilities—compounded, as in Korea and as was once true in Japan, by a tense geostrategic context that invests those vulnerabilities with major political significance. In 1995, Taiwan imported $8.9 billion in crude oil. Unlike Korea, Taiwan lacks substantial coal reserves, so it was also forced to import large amounts of coal. Its combined energy import bill was more than double that of a decade earlier, and more than forty times the level of 1975.[11] APEC forecasts suggest that it could well double again during the coming fifteen years.[12]

Southeast Asia, in contrast to its northern neighbors, presents a more com-plex picture of energy haves and have-nots. As indicated in Table 3-1, Indonesia, Brunei, and Malaysia currently enjoy an energy surplus and export to the rest of the region. Vietnam, as new wells in the South China Sea come into production, is assuming that fortunate position. The Philippines, in the euphoria of recent oil and gas strikes on the Palawan Shelf, eagerly aspires to net exporter status. But Thailand and Singapore remain major oil importers, with little immediate prospect of change. As in South Korea, the consumer revolution, bringing soaring auto ownership in its wake, is increasing energy demand, already stimulated by double-digit industrial growth. Thai oil consumption has risen at the rate of 14 percent annually since 1987.[13]

An ominous but crucial further element in Asia's emerging energy picture is the declining energy surplus of the region's two large traditional exporters, Indonesia and China.

Indonesia has been the largest oil producer in Asia since early in this century; the Sumatra fields of the Dutch East Indies for years fueled the Oriental colonial empires of both Britain and France, not to mention Holland. Indeed, they were among the first targets of the Japanese Imperial Army as it struck southward in 1941. Throughout the post–World War II era, independent Indonesia has continued to be Japan's largest supplier of energy, a standing that not coincidentally has also made it the largest single recipient of Japanese foreign aid for most of the past three decades.

But whether Indonesia can continue its long role as faithful energy supplier to Asia, particularly Japan, is now increasingly in question. Its oil production peaked in 1977, at 82.9 million metric tons, and has since begun a long decline, stimulated both by depletion of the old Sumatran wells and by inadequate incentives to Western multinationals to prospect for oil elsewhere. The geological structure of the Indonesian archipelago, which has created many small pockets of oil, much of it in scattered subterranean basins unusually difficult to prospect and develop, has made this problem of incentives unusually pressing. By 1990, oil production had fallen to less than 85 percent of late-1970s levels, with all signs pointing to further steady decline. According to industry specialists, Indonesia needs to drill about two hundred wells per year to maintain current levels of production. That simply is not happening.

The picture in natural gas is somewhat brighter. Gas, an environmentally attractive fuel, is widely desired in Japan, and a variant is now used in virtually all Tokyo taxis. Indonesian production continues to rise, reaching current levels more than double those of a decade ago. The massive Natuna field, eight hundred miles northeast of Jakarta in the South China Sea, and now being developed in a $40 billion venture between Pertamina and Exxon, contains recoverable reserves of 45 trillion cubic feet, enough to meet Japan's huge current natural-gas demand for a full seventeen years.[14] But this project will not be completed until at least the year 2002, and high levels of carbon dioxide undermine the quality of gas likely to be obtained. Political uncertainties in the South China Sea could also disrupt the project's implementation; China published maps in mid-1995 showing the Natuna fields within its territorial waters, although they lie well over a thousand miles south of the nearest generally recognized Chinese territory.

The explosion of domestic demand, as Indonesians join the automotive age, compounds the problem of steadily declining Indonesian oil production. During the fifteen years after Indonesia hit its high point of oil production in 1977, consumption rose 65 percent, even as production was declining by a fifth, provoking a 41 percent decline in exports.[15] Unless low, subsidized domestic energy prices are sharply realigned, which could be dangerous politically, and major new incentives provided for exploration and production, energy-rich Indonesia will become an oil-importing nation by around the year 2000, forcing Japan and other customers to look elsewhere. Pertamina, the Indonesian national oil company, projects that Indonesia will become a net oil importer sometime between 1998 and 2008, with the precise date depending most crucially on growth of domestic energy demand within Indonesia itself.[16]

China and Asia's Changing Energy Equation

The first thing to remember about China's relationship to the Asian and global energy scene is how *little* energy each Chinese actually consumes today. Chinese consume *far* less per capita of virtually all types of energy than do Americans, Japanese, or South Koreans, and only 40 percent of the world average. They use relatively large amounts of coal, which meets 75 percent of primary energy demands in China, but astonishingly little of anything else. Few of them have personal cars. And appliances such as stand-up refrigerators, dishwashers, and microwaves are virtually unknown. Chinese per capita consumption of oil in 1990 was around one-sixth of the global average, one-tenth of Korea's, one-twentieth of Japan's, and only one-thirtieth of levels in the United States.[17]

China, with its 1.2 billion people—five times the population of the United States and ten times that of Japan—has huge, latent potential demand for energy that will become manifest as its economy develops. Indeed, China's energy policy decisions, such as whether to promote a family car, may well determine the prospects for, and the timing of, another major global oil shock. Those decisions could also profoundly shape the world's environmental future. China today, for example, is already the world's second-largest producer of greenhouse gases, because of its massive use of coal. Its policies to diversify away from that "dirty" fuel or to neutralize its environmental effects could thus likewise have fateful global significance.

Yet these prospects are in the future. Despite China's huge potential domestic energy demand, it remained, from the mid-1970s to the early 1990s, one of the most important energy exporters of Asia. As recently as 1985, the People's Republic of China shipped nearly a quarter of its production abroad. Nations such as Japan had high expectations of China; even in 1990, China exported $2.8 billion in crude oil and petroleum products to Japan alone, the largest amount Japan received from any Asian country other than Indonesia.[18] The PRC also shipped sixteen million tons of coal eastward annually across the East China Sea, mostly to Japanese electric power companies.

Since the early 1990s, China's energy balance has sharply deteriorated, propelled by the dual pressures of double-digit economic growth and transition to a consumer economy. China in 1993 consumed more than two million barrels of oil a day, with its demand for refined oil products rising at close to 20 percent a year.[19] With China's energy consumption per capita remaining little more than one-sixth of Japan's, one-fourth of Taiwan's, and one-third of Korea's,[20] and with explosive growth continuing, its aggregate energy demand will also likely see substantial future expansion.

We can begin to grasp the sobering global implications of rising energy demand in China through international comparisons. It is, for example, quite reasonable to expect that per capita energy consumption in China might reach the level of average Latin American countries shortly after the year 2000. But if that happens, China's total oil consumption could well exceed that of all OECD nations in Europe combined. It could well increase total global oil needs by 20 percent, assuming analogous per capita consumption levels. Similarly, if China reached South Korea's present consumption levels, its total oil consumption would be double that of the entire United States today.[21]

The forces that could provoke such massive and internationally historic increases in Chinese energy consumption are already in motion. Among the most important is economic growth, and the added grassroots buying power that such growth brings. Chinese growth has continued at close to double-digit levels since 1979, and the prospects are good for at least 6 percent annual GNP growth continuing throughout this decade.[22]

China no doubt will continue to be plagued by inequity. But a middle class of 100 to 200 million people (the population of Japan or the United States) is gradually emerging. Within a generation it will have per capita incomes well over $4,000 a year,[23] providing a highly dynamic market for consumer durables.

Over the past fifteen years alone the number of electric fans in China has increased almost twentyfold, and the number of washing machines has risen from virtually zero to 97 million.[24] This consumer revolution, of course, increases China's energy demand sharply. Yet much more fateful changes are now in store.

China's state planners in May 1994 announced plans for a "people's car," an affordable compact sedan.[25] They invited the world's leading automakers to compete to design and manufacture economy models, produced in China. These cars would incorporate such Western technology as modern safety standards, pollution controls, and fuel efficiency, but still retail for under $5,000, making them affordable for China's new middle class.

China's leaders have already declared autos a "pillar industry," meaning that along with telecommunications, computers, and petrochemicals, they will get special government priority. The state is also enticing foreign compact-car manufacturers like Volkswagen, which has produced its Santana in Shanghai since 1985, to cooperate. Within a decade VW has raised its local content to 85 percent, from 190 suppliers in the Shanghai area.[26]

By 1997, Chinese authorities hope to be producing their new people's car on domestic assembly lines modernized and streamlined with foreign technology. By the year 2000, they expect the Chinese auto industry to produce three million passenger cars a year, compared to only 350,000 in 1994.[27] The long-range market: 300 million potential car owners. Given that China already has 1.2 billion people, and that it would have 483 million cars if it had the vehicle density of Germany,[28] this long-term projection seems well within the realm of possibility.

Rush hours in large cities like Shanghai can last from 7:00 A.M. until 9:00 P.M. Given such horribly crowded roads, many Western analysts doubt the practicality of the people's car. But China's leaders are determined, and the concept has obvious—and politically potent—mass appeal. The energy implications are profound, potentially leading China down a path that is ominously parallel to America's. Forty percent of the entire U.S. energy consumption—thirty times China's, on a per capita basis—goes to fuel private cars, and the United States imports half its crude oil as a result.[29]

The problem goes beyond huge prospective increases in overall national energy demand, driven by growth, affluence, and a consumer revolution. The energy supply-demand balance varies sharply across China, with the nation's sizable confirmed onshore reserves located frustratingly far from the sharply growing sources of demand. The most explosive increase of demand flows from the southeastern coastal provinces, notably Fujian and Guangdong. There oil consumption, in particular, is expected to easily double by 2000, as compared to 1990.[30]

The southeast has virtually no local oil production of its own and is heavily reliant on imports. The boom town of Shenzhen in Guangdong, for example, just across the border from Hong Kong, procures over 90 percent of its oil from imports, mainly from Singapore. Its level of imports, like that of most Chinese coastal areas, seems likely to soar over the coming decade.

Most of China's economic heartland, the large cities of the central coast and the industrialized northeast, presently gets its energy from either the coal or the crude oil production of onshore eastern China. Most prominently, there are the famous Taching oil fields of northern Manchuria, long the very symbol of Maoist self-reliance. "Learn from Taching" was a key slogan of the Cultural Revolution. But these eastern fields, which account for 80 percent of China's proven reserves and 90 percent of its current output, are stagnating, like those of Indonesia, after thirty years of intense development.

China appears to have huge reserves of onshore oil in the vast 570,000-square-kilometer Tarim Basin of western Sinkiang, close to the Russian border. Indeed, some experts suggest that Tarim could top Saudi Arabia's proven reserves of nearly 250 billion barrels.[31] Without question its huge underground energy reservoir should exceed total current proven reserves from the rest of China combined.

Prospecting conditions, however, are torturous, with seismic exploration teams battling sand dunes thirty or forty feet high, rivaling those of Saudi Arabia's Empty Quarter, the Rub al-Khali. And the oil, once located, is devilishly difficult to extract. Tarim's wells are the deepest in the world—on average twice the depth of Saudi Arabia's, and triple those in Manchuria's Taching fields. The geological formations at Tarim are also very old, resulting in faults that fragment deposits into frustratingly small, hard-to-find pockets of oil. All this makes wells in the Tarim Basin exorbitantly expensive to drill—when the oil can be located at all. Advanced Western technology would definitely help.

Tarim is also so remote that huge infrastructural investments in pipelines, roads, and telecommunications are inevitable. The price tag on a single proposed 3,500-kilometer pipeline alone, to transport the oil from Sinkiang across deserts and mountains to Henan province on China's east coast, could run $10 to $12 billion. In 1992, all Sinkiang accounted for less than 7 percent of China's oil production, and this situation may not change substantially until well into the next century. The International Energy Agency suggests, for example, that without foreign participation the Tarim Basin will probably not be developed before 2010.[32]

Some analysts, particularly Chinese, believe the ambitious Three Gorges Dam project, in central China, now in its infant stages, will sharply reduce China's reliance on the broader world for energy imports. This stupendous undertaking, proposed seventy-five years ago by Sun Yat-sen, would involve building a 1.2-mile, 5,150-foot-high stretch of concrete across China's largest river, the turbulent Yangtze.

The project could require as much as $77 billion in investment capital over its life, and take twenty years to complete.[33] It would generate as much energy as ten nuclear power plants, clearly reducing China's chronic energy shortage, and prevent the periodic flooding of the Yangtze that has claimed 500,000 lives this century. Three Gorges would, however, also submerge 28,000 acres of farmland and twenty towns, driving 1.4 million people from their homes, and cause incalculable environmental damage.

Three Gorges thus faces gigantic obstacles that have already provoked the World Bank to declare it economically unviable. Its drawbacks have likewise inhibited the United States, Japan, Germany, and Canada from offering the active government support crucial to raising large amounts of foreign investment capital. The project's current major source of finance is thus a 2 percent tax on all electricity generated in China, the proceeds of which will go to dam construction.

The Three Gorges project will require huge resources and take a generation to complete. It is thus extraordinarily vulnerable to the sudden changes of politics and policy that have characterized China throughout the twentieth century. Even if the dam is ultimately built it can do little to remedy China's grave energy situation during the critical coming ten to twenty years.

The prospect for offshore oil within internationally recognized Chinese waters also looks poor, at least for the foreseeable future. Multinational firms have already invested over $2 billion—with discouragingly poor results, despite high early expectations—in the East and South China seas off China, as well as the Bohai Gulf. Amoco in April 1993 announced a $650 million development program for the Liuhua 11-1 deepwater oil field in the South China Sea, 120 miles south-southwest of Hong Kong.[34] Yet broader prospects for at least the portions of the adjacent seas in clearly Chinese territorial waters remain clouded.

Some analysts see significant potential in parts of the East and South China seas where the claims of China and other nations overlap.[35] Among the most promising areas are the great offshore basins in the East China Sea, southeast of Shanghai and northwest of Okinawa. There the mighty Yellow and Yangtze rivers deposit sediments drained from nearly one-third of all China.

The Yellow River, in particular, has been notorious for centuries. It tears topsoil from treeless hillsides, spiriting it away, in the great muddy torrents that give the river its name, across three thousand miles of China to the ocean. Together the Yellow River and the Yangtze deposit nearly three billion tons of sediment every year into the East China Sea, there to mature into hydrocarbons over untold millennia.[36]

The bulk of this sediment, now close to a mile thick in many places, lies nominally in international waters, on the continental shelf beyond the river deltas.

Yet China, not surprisingly, considers it patrimony. Disputed sections of the South China Sea also reportedly have oil potential, both near the Spratly Islands, claimed by six nations in addition to China, and in deep waters (more than two and a half miles deep in some places) to the south of Hainan.[37] But as in the Tarim Basin and Indonesia, complex geological structures make it difficult to calculate just how cost-effective actual exploitation might be.

The uncertainties regarding China's energy potential leave the poverty-stricken nations of Indochina to narrow Asia's looming petroleum gap. But they cannot do so by much. Although Vietnam increased its oil production almost tenfold between 1989 and 1994, to some 167,000 barrels a day, it remains a dwarf among producers. Even the end of the American trade embargo against Vietnam and the return of technologically sophisticated U.S. producers and explorers like Mobil Oil and Halliburton cannot make a decisive difference. And Chinese military activities in the South China Sea inhibit Vietnamese production still further.

Asia's Rising Share of World Energy Demand

In 1991 the Asia-Pacific countries as a whole consumed 23 percent of the world's energy, or the equivalent of 35 million barrels of oil a day. A decade earlier their share had been only 18 percent.[38] In 1990, Asia-Pacific regional oil demand overtook that of Western Europe, making the region the second-largest oil-consuming region in the world after the United States. Demand has been growing steadily ever since. As East Asian demand was so dynamically rising, European and North American energy consumption hardly grew at all. That in Africa, Eastern Europe, and parts of Latin America, caught in the throes of recession and debt crisis, was actually declining.

Driven by explosive economic growth, East Asia's overall energy demand should continue to rise, making it the world's growth market for oil, liquefied natural gas (LNG), coal, and other forms of fuel. The International Energy Agency suggests that overall growth in East Asian energy use could average well over 4 percent to the year 2010—roughly double the rate of increase that IEA anticipates for energy demand worldwide.[39] Where will the oil to supply Asia's burgeoning demand come from if not from Indonesia and China, the traditional suppliers? What alternate energy substitutions—solar, geothermal, LNG, coal, or nuclear—will be made? Such questions will become ever more pressing as the years go by.

East Asian energy demand, especially that outside China, focuses to an unusual degree on oil. It accounts for 51 percent of total regional energy consumption, compared to the global ratio of only 40 percent. A heavy concentration of energy-intensive manufacturing (steel and petrochemicals, for example), oil-fired electrical generation, and high, rising automobile demand help explain these high oil-consumption ratios. Coupled with the Asian region's extraordinary rates of overall economic growth, this heavy dependence on oil leads, not surprisingly, to soaring demand for oil. While overall world oil demand was essentially flat in 1993, East Asia served as a decisive engine of growth, with regional petroleum product demand rising 4 percent. Outside Japan, oil demand in 1993 rose 6.8 percent, with growth reaching 11.4 percent in South Korea, as well as 8–9 percent in both China and Thailand.

These buoyant levels of growth in oil demand could well continue, with only modest alteration, for another generation. Indeed, Asia-Pacific oil demand could well rise by over seven million barrels over the balance of this decade alone.[40] ASEAN and China's coastal provinces will likely show an especially rapid rate of increase. In all, the Asia-Pacific region, which currently consumes around a quarter of the world's energy, could account for 40 percent of the increase in global energy demand over the coming fifteen years.

The growth of energy-intensive industries crucial to infrastructural development, such as steel and petrochemicals, will be one important source of new demand for energy, especially along the Chinese coast. These are industries on which the most rapidly growing lower-income Asian NIEs, especially China, now place considerable priority. But the largest increments to Asian energy demand over the coming generation will flow from the transportation sector, which is expected to account for over half of East Asian consumption by 2010.

East Asia's exploding auto market will lead the way. Throughout the first half of the 1990s, in economies growing at close to 10 percent a year, Chinese and Southeast Asian vehicle sales rose at a clip of 30 percent annually, with no slowdown in sight.[41] Yet every one of the developing East and Southeast Asian nations still has less than half the ratio of passenger vehicles to population now prevailing in Japan, with Thailand, Indonesia, and the Philippines having only one-sixth the ratio of cars.[42] All of these countries have more room to drive than does crowded Japan, particularly once cities like Bangkok, Shanghai, and Taipei have dealt with their massive pollution and infrastructure problems. And as their middle classes steadily expand, more and more of their peoples will be in the auto market. China's auto-related additions to regional oil demand, triggered by the "people's car," could be especially fateful, as noted above.

Air travel will be another major stimulus to growth in energy demand. East Asia still has well over a billion people who have never boarded an airplane. Boeing believes that the region will provide more than half of global demand for new aircraft over the next generation, and the overcrowded Asian lines are a major market for used planes as well. Air travel grew in China at an annual rate of 20 percent continuously between 1978 and 1992, yet still remains less than 6 percent of total transport, very low for a nation of China's size.[43] Aviation fuel demand should grow at at least double the rates of overall energy usage across the coming two decades.

Rising Competition for Supply

Asia's energy demand, centering on oil, will not only rise. It will also, as has been suggested earlier, progressively broaden, to include not only Japan, but a wide range of even more rapidly growing nearby economies. Recent APEC forecasts, presented in Table 3-2, suggest that within fifteen years there will be four or five major competitors for existing Asian oil supplies in regional markets traditionally dominated by Japan alone.

By the year 2010, should current forecasts prevail, Japan's share of Asian oil imports will have fallen by half, to a bit more than one-third of the region's total. Greater China (including Taiwan and Hong Kong) will account for more than a quarter, with mainland China's share rising sharply. Korea and ASEAN—both with still-rising requirements—will have to fight for the rest, in potentially tight regional and global markets. The intensity and complexity of emerging multipolar energy-supply rivalries in Asia are developments for which the region (and the world) remain remarkably unprepared.

Filling the Gap

Natural gas, as yet a remarkably underappreciated fuel source in the region, may provide some limited relief from the increasingly serious energy shortages that loom before high-growth East Asia, especially because it is extremely clean-burning and environmentally acceptable. Natural gas, for example, accounts for only 10 percent of total energy consumption in Japan, compared to 20–25 percent in the United States and Europe.[44] Production of LNG, in contrast to oil at least, has kept pace with demand, the bulk of which has flowed from crowded, heavily industrialized Japan, Taiwan, and South Korea, where environmental concerns are strong. Indonesia, the East Asian region's largest producer, has a substantial share of proven world LNG reserves. Development of the huge Natuna fields in the South China Sea promises to extend the active life of those Indonesian reserves well into the early twenty-first century,[45] provided that political complications do not intervene. Malaysia, Australia, China, and Brunei, within the western Pacific, also have significant reserves, leading APEC forecasters to sug-

Table 3-2

EMERGING ASIAN OIL IMPORT RIVALRY?

Importing Nation	Share of Total Asian Oil Imports (%)		
	1992	2000	2010
Japan	77.4	53.2	36.5
China	—	10.9	19.4
Taiwan/Hong Kong	10.0	9.7	9.0
Korea	21.0	20.3	18.3
ASEAN	—	5.9	16.9

Source: APEC International Advisory Committee for Energy Intermediate Report, June 1, 1995.
Notes: Based on 1995 national energy supply-demand projections, in metric tons of oil equivalent (Mtoe), for Japan, China, Taiwan/Hong Kong, South Korea, and ASEAN; 8.4 percent of net 1992 imports into the nations listed were from China and ASEAN.

gest that LNG could supply as much as a tenth of Asia's total energy demand by the year 2010.[46]

Yet the incentives for the huge and risky investments in LNG export facilities like pipelines and liquefaction plants, without which LNG cannot move internationally to potential consumers, remain limited by low fuel prices on the world market. For LNG to become more attractive, oil prices will have to go up and stay up, in predictable fashion. And LNG facilities are always developed jointly by producer and consumer. No consumer likes higher prices.

Some analysts suggest that Russia could provide an attractive source of both oil and natural gas for East Asia.[47] Certainly it has the resources to do so, with an estimated 20 percent of all the oil and 35 percent of all the gas on earth lying within its borders.[48] Gas from the Yakutia region of Siberia could be shipped south through North Korea to supply South Korea, while that from Sakhalin could be shipped by pipeline and tanker south to both Korea and Japan.[49]

Unfortunately, however, there are myriad economic and political complications with these massive prospective projects. The problems are intensified, especially in the case of natural gas, by the huge scale, and associated financial risk, of constructing related pipelines, ports, liquefaction facilities, and other such infrastructure. Indonesia's new Natuna fields in the South China Sea, with projected development costs of nearly $40 billion, present different difficulties, lying as they do at the edges of China's huge maritime territorial claims. The necessary pipeline to transmit gas from Yakutia south to Korea faces parallel political complications, crossing multiple unstable jurisdictions in the Arc of Crisis, including the Russian Far East and North Korea. It would cost at least $5 billion to build, while an extension down to Southeast Asia, connecting to Japan's proposed $10 billion domestic pipeline system, could cost $25 billion, and likewise provoke political complications.

Apart from the cost of infrastructural investments, existing oil and gas wells in Russia are plagued by difficulties in servicing and maintenance that are causing steady production declines throughout the country. Oil production in Russia peaked in 1987 at around 11.4 million barrels/day and has been declining ever since.[50] By the end of 1993, it had collapsed to seven million barrels per day and was still declining, plagued by maturity, poor maintenance, and lack of investment in the older West Siberian fields.[51] This astonishing drop is equal to 60 percent of total U.S. oil production and is greater than the output of any OPEC state except Saudi Arabia.[52] Prospects for an early recovery are bleak, with Russian production expected to fall to as little as one-third of peak levels by the late 1990s. At least $100 billion of new investment in Russia's abused and depleted oil fields would be needed even to return output to the levels of a decade ago.[53]

Coal is another alternative to East Asia's heavy oil dependence. For China, in particular, coal already is and will continue to be a vital energy source; it currently provides three-quarters of China's energy, and China is the world's largest consumer of coal. That country's massive trillion tons of proven coal reserves are also the third-largest in the world, while the Asia-Pacific region as a whole has a comfortable 29 percent of global coal reserves.[54]

Like nuclear power, however, coal poses major environmental dangers. It makes China—still a middle-sized economy in global terms—the second-largest producer of greenhouse gases on earth, amid forecasts that it could more than double its CO_2 emissions within the coming fifteen years, and become the world's number one CO_2 polluter by 2020.[55] Already, acid rain from China's wholesale use of coal is beginning to defoliate forests in Toyama and Shimane prefectures, far across the Sea of Japan. On top of other drawbacks, coal cannot accommodate China's coming auto revolution. In the end the Middle Kingdom, and Asia with it, is forced back onto oil, coming down hard again on its painful Achilles' heel.

One option, of course, could be offshore oil produced within the region. Asia's seabeds, particularly those close to China, are covered with heavy layers of sediment rich with hydrocarbons, deposited over millennia by the region's muddy, silt-laden rivers. The region has an abnormally wide continental shelf, more accessible to drilling than the deeper sea, with over 5 percent of the world's shelf adjacent to China alone. Though offshore oil may perhaps be there in major quantities, it is unlikely to be exploited within this decade in a sufficiently systematic way to prevent fatefully rising Asian energy dependence on other parts of the globe. Much of the oil is in rather deep seas and far from the coast. Much of it, especially in the East China Sea, appears in small, frustratingly hard-to-develop pockets. And much of it is also in politically contested waters, where political risk factors inhibit exploration.

The technology of offshore exploration and production, to be sure, is improving rapidly. For example, in the Gulf of Mexico and offshore Brazil, Shell and other Western oil producers are already using deep-diving robots for construction and maintenance; they service huge platforms tethered by steel tendons to barn-sized anchors on the ocean floor.[56] Yet a maze of complications, many of them political, make it unlikely that Asia can possibly slake its voracious thirst for oil internally for many years to come, despite the rapid advance of offshore exploration, drilling, and production technology.

East Asia as Rising Oil Importer

Currently Asia meets roughly half of its oil needs through imports from outside the region. And as demand rises, spurred by hypergrowth and the consumer revolution, this could climb to two-thirds by the year 2000.[57] If imports from Russia remain constrained by country risk and huge short-run infrastructural requirements, and offshore prospects in the China Seas continue to be clouded by political uncertainty, Asia cannot avoid deepening its dependence on the low-cost Middle East. There wellhead production costs range as low as $1–2 a barrel, and substantial excess production capacity still remains. Indeed, according to East-West Center estimates, by the year 2000, 87 percent of all the oil East Asian nations import will flow from the Middle East, up from 70 percent at present. By the year 2010, import dependency on that volatile region will rise to 95 percent.[58]

Should this projected pattern indeed materialize, a growing fleet of heavily laden supertankers will plow east across the Arabian Sea and Indian Ocean in coming decades, headed for Singapore, Hong Kong, Shanghai, Pusan, and Yokohama. As Table 3-3 suggests, East Asian oil imports from the Middle East could well triple in the next fifteen years alone, to a very substantial share of total world oil consumption.[59] Unless noneconomic forces intervene to arrest this trend, the two major economic and geopolitical centers of the non-Western world would be joined in an unprecedented embrace, with global implications.

Japan and, to a lesser degree, South Korea and Taiwan are already heavily dependent on the Middle East for oil imports. Indeed, Japan in 1993 imported

Table 3-3

THE DEEPENING EAST ASIA–MIDDLE EAST ENERGY LINKAGE (1993–2010)

A. *The Asia-Pacific Crude Oil Equation*

	1993	2000	2010
Oil demand, million bbl/day	14.9	19.8	26.6
Oil supply, million bbl/day	6.9	6.9	6.7
IMPORTS	8.0	12.9	19.9

B. *Rising Dependence on the Middle East*

Share of imports from Middle East, percent	70	87	95
Volume of imports from Middle East, million bbl/day	5.6	11.2	18.0

Source: Fereidun Fesharaki, Allen L. Clark, and Duangjai Intarapravich, eds., *Pacific Energy Outlook: Strategy and Policy Imperatives to 2010*, East-West Center Occasional Paper, Energy and Minerals Series, No. 1 (Honolulu, March 1995).

around three-quarters of its entire oil supply—or about 70 percent of its entire consumption—from that volatile area. What is changing—a matter of paramount economic and security importance for both the region and the world—is the entry of China into this whole equation, within the context of broadening energy competition that also includes ASEAN as a major importer.

Until the early 1990s, Chinese oil imports were insignificant—less than three million metric tons annually. But from 1990 they began to steadily escalate. Even though China continues to export some oil, its imports have surged, because of the location of existing oil fields and refineries coupled with the difficulties of internal distribution. In 1993 alone those imports grew by 37 percent, to 309,000 barrels per day.[60] Fueled by explosive growth in coastal areas, especially those of the southeastern coastal provinces such as Guangdong and Fujian, China became a net oil importer during the fourth quarter of 1993, for the first time since the mid-1960s. And a significant share of its emerging imports were from the Middle East.

With Chinese internal forecasts suggesting domestic oil demand of 175 million tons by the year 2000, and production capacity of little more than 145 million,[61] the prospects are clearly strong for large and rising Chinese oil imports. Even the relatively conservative International Energy Agency suggests that these imports will likely rise to 900,000 barrels per day by the year 2000, and to roughly 2.8 million bbl/day, or triple that volume, by the year 2010.[62] Including Taiwan and Hong Kong, Greater China's oil imports will likely approach those of Japan within a generation.

Currently around 43 percent of China's crude oil imports are from the Middle East, with Oman as the largest supplier.[63] Given the low cost of Persian Gulf oil at the wellhead, however, that unstable region will be forced to meet the overwhelming proportion not only of China's but of all East Asia's new incremental demand, as suggested in Table 3-3. More than ten million barrels of oil a day could be flowing by supertanker east to Asia, with an increasing focus on China, as the twenty-first century dawns.

How will China's growing future oil supply flow from the Persian Gulf to the local markets all along its thirsty, growing coastline? Who will ensure that such supplies are not interrupted? Will China build a blue-water navy to provide its own security for oil flows to the Gulf, or can it, as Japan and South Korea have done, rely on the United States to guarantee passage through the Strait of Hormuz and beyond? If China chooses the former course and expands its naval forces accordingly, how will the other nations of East Asia react? What form of cooperation will China extend to Middle Eastern states, including pariah states like Iran and Iraq, in return for ever larger, and ever more urgently needed, quantities of oil? Will energy ultimately seal an Islamic-Confucian entente that could decisively shift the balance of global power away from the West? Historic energy-linked issues surely loom on East Asia's horizon, with fateful implications for all the world.

ASIA AND THE NUCLEAR THRESHOLD

Late summer is the Buddhist season of Obon in Japan, when for hundreds of years mourners have gathered to meet the returning spirits of the departed. Just after 8:00 A.M. every August 6, very much in the Obon tradition, Japan's NHK national television network, with a viewing audience of several million, suspends its regular program and turns from current news and morning variety shows to a retrospective on the past. The programming shifts to a humid, middling-sized town on the shores of the Inland Sea without mountains or other striking visual features. Were it not for a few split seconds of infamy half a century ago, Hiroshima would never inspire the attention of the world.

As NHK shifts to the Hiroshima scene, its cameras pan an impressive but aging assemblage from throughout the globe, dressed in somber hues, gathered before an eternal memorial flame to commemorate the victims of history's first nuclear bombardment. After closing for moments on the eternal flame, cameras also linger on the gnarled, rusting girders and cracked foundations of Hiroshima's Industrial Exhibition Hall, only meters from the ground zero of 1945, and its mute yet enduring testimony to the ferocity of the original attack. After minutes of discordant, atonal music, designed to symbolize the advance of the American bomber *Enola Gay* toward Hiroshima, there is a minute of silence, precisely at 8:15, and memorial tributes from throughout the land. Around 8:30, the commemoration ends with the release of hundreds of doves of eternal peace, and NHK's programming returns to the mundane world of morning soap opera and children's shows.

Nuclear power plays an anomalous and contradictory role in Japan, as indeed in all Asia. On the one hand, it is the origin of unique tragedy that no other part of the world has ever actually experienced—nuclear attack. On the other, it is the prospective source of unique benefit—nuclear power provides an especially attractive answer to many of Asia's pronounced energy needs. As the region's economy grows, it ironically becomes more and more dependent on the same atom that has caused such suffering in the past. And that dependence threatens to open a Pandora's box of both economic and security dilemmas for the future.

The Special Attractions of Nuclear Power in Northeast Asia

Nuclear power has, as an energy source, important attractions for much of Asia, especially the Northeast Asian Arc of Crisis and Japan. For nations chronically short of energy, it has the merit of yielding it in large amounts, without bulky imports. In the extreme form of the fast breeder reactor, in fact, nuclear power actually becomes a self-renewing source of energy that liberates longtime energy importers from their persistent dependency. Transportation costs are low, thanks to nuclear fuel's compactness. And the marginal costs of producing nuclear power are also low, if some way of coping with high startup costs, waste products, and safety concerns can be found.

For countries concerned with acid rain, global warming, and other forms of air pollution, nuclear power offers a clearly attractive substitute for coal-fired power generation that emits no greenhouse gases. Provided proper safety measures can be maintained, nuclear power is an environmentally friendly substitute for oil and coal for many East Asians.

With its broad range of economic attractions, many unique to the region, nuclear power has become quite popular in Northeast Asia. By the end of 1992, Japan received 28 percent of its electricity supply from nuclear power, compared to 20 percent in the United States.[1] In South Korea, nuclear plants operating in four giant nuclear clusters around the country that already house nine large power stations provide roughly 40 percent of total electricity.[2]

In contrast to the United States and most of Western Europe, Asia, particularly its northeast quadrant, intends to rely even more heavily on nuclear power in future than it does at present. Currently the region represents roughly 14 percent of global installed nuclear capacity. But U.S. Department of Energy forecasts suggest that Asia could well provide 48 percent of the entire increase in nuclear capacity throughout the world between 1992 and the year 2010.[3] At least $160 billion in capital spending on nuclear plants is scheduled for the next decade alone.[4]

The bulk of this increase—a total of 32.6 gigawatts—would come in Japan and Korea, making them two of the three nations with the largest-capacity additions on earth over the next generation, as Table 4-1 indicates. China and Taiwan are projected to be sixth and eighth respectively. The United States, by contrast, is expected to retrench. The Department of Energy expects America to have close to 10 percent *less* nuclear capacity in 2010 than it does now.

Relative to its size, Korea has particularly ambitious civilian nuclear plans, which transcend the 38th parallel. South Korea plans to nearly double its existing capacity to as much as 18.4 net gigawatts by the year 2010—as much as the existing capacity of Russia today, which is around 18 net gigawatts. This would involve the building of at least eighteen new reactors at a cost of over $20 billion.[5] The Korean Energy Development Organization (KEDO), established by the United States, Japan, and South Korea, would, if the October 1994 U.S.–North Korean nuclear agreement is fully implemented, also build light-water reactors with two net gigawatts of capacity for North Korea, at a cost of around $5 billion. In return, Pyongyang would agree to scrap its military nuclear program.

Together the North and South will thus more than double their nuclear capacity over the coming fifteen years, by even conservative estimates. And if North Korea's economic opening to the world proceeds smoothly, they could need still more. Household use of electricity is virtually nil in the North at present, and industry is stagnant; changes on both fronts would trigger a surge of local demand north of the 38th parallel, if only new power-generation facilities could be paid for.

China also has ambitious nuclear plans, in which Korea hopes to participate actively. Confronting both chronic and frequent brownouts in the face of explosive, sustained growth, together with an ever more polluted environment because of widespread use of coal, China is placing high priority on nuclear power, especially along its energy-short southeastern coast. As China plans to build forty atomic power plants by 2020 and one hundred by 2050, its market for nuclear power equipment, much of which Hanjung and other Korean companies hope to supply, is expected to expand to a $200 billion scale by early in the twenty-first century.[6] Such growth could well accentuate both China's and Korea's nuclear orientation, as well as technical cooperation between them.

Taiwan is also a major user of nuclear power, generating more than a third of its electricity (33.5 percent) that way in 1993.[7] In 1994 it had three nuclear plants with six reactors, the fourth plant having been suspended in 1986 following

Table 4-1

THE GROWTH OF NUCLEAR CAPACITY, 1992–2010: EAST ASIA'S KEY ROLE

Country	1992 Capacity	Increment by 2010	
		Lower Reference Case	Higher Reference Case
1. Japan	34.2	+11.5	+22.5
2. France	57.7	+4.4	+12.1
3. Korea (North and South)	7.2	+7.7	+11.2
4. Canada	14.6	+1.8	+8.3
5. Russia	17.8	+1.1	+7.5
6. China	0.3	+2.4	+3.9
7. Brazil	0.6	+2.5	+3.3
8. Taiwan	4.9	+1.9	+2.9
U.S.	99.0	−8.3	−8.3
Britain	11.9	−2.0	−3.9

Source: U.S. Department of Energy, *International Energy Outlook*, 1994 ed., p. 37.
Notes: Figures are for projected operable nuclear capacity, expressed in net gigawatts. Korean figures adjusted to take account of October 1994 U.S.–Korean nuclear capacity by 1.1 net gigawatts beyond previous projections.

public protests sparked by the Chernobyl disaster in the Soviet Union. Yet despite somewhat more popular resistance than elsewhere in East Asia, the economic and national-security logic of nuclear power seems persuasive to Taiwanese planners as well. They intend to expand local nuclear capacity by a full two-thirds over the coming fifteen years.

There Is Only One Atom

Civilian nuclear power appears quite innocuous in military terms. Indeed, many nations with extensive civilian nuclear capacity, particularly in East Asia, actually lack nuclear weapons. But when a country develops a civilian nuclear capability, it also proceeds much of the way toward possessing a nuclear device. There are not, after all, two atoms, one civilian and one military. They are ultimately the same atom.

Civilian and military uses of nuclear energy are linked at two critical points in the nuclear fuel cycle. First, when it is mined, uranium is not quite suitable for nuclear reactors; it must be "enriched" to about 3 percent U-235 to sustain a controlled chain reaction. The capacity to produce such low-enriched uranium (LEU), to be sure, does not automatically create the capacity to produce the highly enriched uranium (HEU) suitable for use as the explosive core of atomic bombs like the one used on Hiroshima. Yet that capacity is only a limited, albeit fateful, technical step beyond.[8]

A second critical point in the fuel cycle comes after the enriched uranium has been consumed. Nuclear waste or ash remains, some of which can be chemically separated, or "reprocessed," and used again as fuel. This "reprocessing" generates plutonium, the substance of the Nagasaki bomb.

Plutonium has two uses. It can be separated for innocuous use as civilian reactor fuel. But it can also become one of the two basic materials needed in creating an atomic bomb.

Even standard reactors can generate considerable plutonium—as much as two hundred kilograms a year. Crude implosion bombs require as little as ten kilograms. An operating nuclear reactor can thus produce enough raw material for the explosive core of a bomb every two or three weeks.

Ample supplies of plutonium or enriched uranium are not enough by themselves, of course, to make a bomb. Yet they are an important precondition. And both deadly substances will become plentiful across Northeast Asia, especially Japan, in coming years.

Plutonium and the Nuclear Fuel Cycle

The heart of the dual ability to generate "raw material" for nuclear weapons while efficiently producing civilian nuclear fuel is what is called a complete nuclear fuel cycle program. Japan's variant, conceived in the shadow of oil shock during the mid-1970s, will generate if fully implemented around eighty to ninety tons of mixed-oxide plutonium there by the year 2010, through a combination of imports from British and French nuclear reprocessing centers (thirty tons) and domestic reprocessing of nuclear waste from conventional uranium-based reactors (fifty tons). Although this material will not technically be weapons-grade plutonium, the separation process required to convert it into the raw material for bombs would be a relatively simple one.

Over the longer term, the accumulation of plutonium in Japan will be even more substantial as the nation shifts from currently conventional light-water reactors to the technically more efficient fast-breeder reactor (FBR). Fast-breeder reactors such as Japan—alone among major industrialized nations—is now developing have the potential to squeeze about sixty times more energy from uranium fuel than the light-water reactors do. Yet while burning a refined version of the plutonium given off as a by-product in LWRs, they also breed more plutonium.

Japan's Atomic Energy Commission projects that the plutonium generated in the nuclear fuel cycle by 2010 will be consumed through a projected demand of eighty tons from conventional reactors. Yet many observers believe consumption could be significantly less. Japan's Citizens' Nuclear Information Center projects demand at only twenty-three tons, and consensus estimates are around forty.[9] Eight kilograms of plutonium are enough to make an atomic bomb with the capacity of that dropped on Nagasaki in 1945.

Japan's quiet accumulation of plutonium has already begun. It commenced with a small power-industry shipment from Europe in 1984, escorted by the U.S. and French navies.[10] The second step was taken in January 1993, with the secret fifty-nine-day maritime shipment of 1.7 tons of plutonium (enough for 120 crude nuclear weapons) halfway around the world from the French port of Cherbourg aboard the reconverted British freighter *Pacific Crane*, rechristened *Akatsuki Maru* ("Ship of Dawn").

Taking the utmost precautions to protect the small Coast Guard Agency convoy (only one lightly armed ship) that accompanied the plutonium supplies, and fearing diplomatic incidents, the Japanese government operated with deepest secrecy, ordering the *Akatsuki Maru* to take a furtive, circuitous route around Africa. From there it and its fateful cargo, loosely powdered and stashed in protective casks, proceeded across the most remote recesses of the Indian Ocean south of Australia, and then across Melanesia to Japan. Sixty-nine warships, large and small, were deployed as a protective cordon when it entered Japanese waters in early January 1993. This voyage was only the first of many projected transoceanic shipments that will take place over the coming twenty years, as Japan slowly puts its vast, distinctive, and in some ways paradoxical nuclear-fuel-cycle program in place.

To be sure, the long-term energy economics that East Asia confronts favor both reprocessing and ultimately fast-breeder reactors as well. Yet they are far less compelling in the short run, especially for Japan, than appeared likely in the 1960s and 1970s, when the original programs were conceived, in the shadow of the oil shocks. Energy prices are lower, and demand is rising less rapidly than projected. Safety issues, in the wake of Chernobyl and Three Mile Island, are more compelling, and a huge new post–Cold War plutonium fuel supply has emerged in the form of superfluous nuclear weapons.

Numerous expensive steps have already been taken that render the fuel-cycle program virtually irreversible, however much its implementation schedule may be modified in the light of changing circumstances and second thoughts. Ground has already been broken on a huge $10 billion reprocessing complex near the village of Rokkashō, at the northern end of Honshu. This thousand-acre site, scheduled to become operational just after the year 2000, houses facilities both to produce enriched uranium and also to store and reprocess fuel to extract plutonium. There is also a dump for low-level nuclear waste.

Japan's first prototype FBR, capable of generating fissionable plutonium, was completed in Fukui prefecture, on the Japan Sea coast opposite Korea, in the fall of 1991.[11] Japan's first actual electricity-generating breeder reactor, a $6 billion complex named Monju after a Buddhist god of light, reached criticality on the Tsuruga peninsula, 250 miles west of Tokyo, in April 1994. It is funded not only by government but 20 percent also by Japan's nine electric utility companies. They would take a huge capital loss if Japan's plutonium-generating nuclear fuel-cycle plans were reversed.

Both the utilities and much of the bureaucracy have grown increasingly am-bivalent, as both international criticism and the costs of the FBR program, now well over $20 billion, have mounted. The devastating Kobe earthquake of early 1995, with its epicenter little more than a hundred miles from the Monju FBR facility, stirred further misgivings, although no direct damage to Monju was done.[12] But the huge costs already absorbed, the reality of the original commit-ment, and the long-term logic for energy-short Japan have all kept the program rolling slowly forward, against global trends prevailing everywhere but in East Asia. Both the United States and France, for example, have already abandoned once-ambitious FBR programs of their own.

Around the year 2010, according to current plans, Japan will have three breeder reactors in operation, including Monju. These will have generated fifty to sixty tons of plutonium domestically. Together with the thirty tons that Japan expects to import from Britain and France before its own breeder reactors and reprocessing plants are fully in operation, it will have amassed in all close to one hundred tons of plutonium. This is more than the amount currently contained in all the nuclear warheads of both the United States and the former Soviet Union.

The official rationale in Japan for this massive accumulation of plutonium over the next twenty years is that a plutonium-cycle nuclear program is both a highly efficient way to generate energy and also uniquely suited to the energy requirements of Japan in coming years. Most important, it is a "semidomestic" form of energy, since it is self-renewing, and supplants large quantities of imports. "By the year 2010 or 2015," the deputy director of the Monju breeder reactor construction project, Kobori Tetsuo, has suggested, "we will have big international competition, maybe war, over energy."[13] In such a situation, it would clearly be improvident for Japan to grow any more deeply dependent on imported energy than necessary.

The rationale for breeder reactors, in the view of Japan's industrial strategists, has been compelling, given the high-cost-energy future that they have forseen. Fast-breeder reactors, which generate plutonium, are much more energy-efficient than conventional uranium-based reactors, as noted above.[14] Without the recycling of which they are capable, so the logic has gone, Japan by the year 2010 would be consuming 10 to 20 percent of the world's entire uranium production, given its ambitious civilian nuclear program. Without domestic uranium supplies, it would be shackled into the same sort of import dependence in the nuclear area that it currently suffers with respect to oil, in a world where all forms of energy will likely be in short supply.

The construction costs of fast-breeder reactors are huge—four to six times those of conventional uranium reactors, in the calculation of even Japanese specialists.[15] In addition, the FBR program has an extraordinarily long lead time— FBRs could not contribute significantly to Japan's electrical grid for at least forty years,[16] or until after the year 2030. Transportation costs for plutonium from reprocessing centers in Europe will also be extremely high, during the period (roughly the next fifteen years) when Japan will be actively importing the deadly substance. Professional estimates suggest that the early-1993 plutonium ship ments from Europe, for example, cost around $185 million to transport, in view of the long distance and the extraordinary security precautions required—significantly more than the plutonium itself was worth.[17]

The most disquieting implication of Japan's nuclear-cycle program and the massive plutonium stockpile that it would generate is the potential impact on neighboring nations of the Northeast Asian Arc of Crisis, unless further thought is given to multilateral mechanisms for constraining proliferation and providing for responsible storage of nuclear waste. Most of these nations have parallel energy vulnerabilities and a parallel desire to neutralize them in a cost-effective way. They also have considerably more volatile domestic political prospects than Japan, more threatening military deployments, and less reliable systems for monitoring and storing nuclear by-products.

The most conspicuous case is secretive and unpredictable North Korea. Like Japan, it has few domestic energy resources. And it has virtually no capacity to handle a major increment in electric-power demand, when and if its benighted economy ever begins to grow. North Korea could, like Japan, benefit from a plutonium-generating FBR program, quite apart from any aspirations for nuclear weapons that it may have. Indeed, Pyongyang's argument for building its suspicious Yongbyon nuclear complex in the early 1990s was a disarmingly innocent one: simply preparation for a fast-breeder program like peaceful Japan's.

South Korea has also pressed the United States for the right to recover plutonium from U.S.–supplied nuclear fuel, just as the Japanese are already doing.[18] It has also gone ahead unilaterally with an elaborate fast-breeder reactor research program, about which the U.S. government has expressed clear, formal reservations.[19] Under this program, Seoul plans by the year 2011 to build a nuclear fuel-cycle plant roughly similar to the one Japan opened at Monju in 1994.[20]

One major factor inevitably entering into long-term Korean nuclear calculations—in both North and South—must be the possible effects, in both energy and security terms, of a prospective reunification. Whatever short-term dislocations might occur, reunification would almost certainly produce a surge in long-term energy demand, especially in the heretofore deprived North, which no part of Korea has the domestic capacity to supply. Nuclear power, including FBRs, looks better to Koreans than deepening dependence on what could be an energy-short outside world. On the security side, some residual nuclear capacity gives insurance, in Korean eyes, against big-power manipulation, including the prospect of a militarily resurgent Japan.

Apart from the long-term energy, security, and efficiency benefits that Japan also sees, the Koreans view reprocessing and FBRs as a solution to the severe nuclear waste management problem now emerging in their country. South Korea's nine reactors have already produced some 43,700 drums of low-grade radioactive waste. Its current temporary storage facilities are nearly two-thirds full, it is building seven more reactors, and it has no permanent place to store its nuclear waste. Rather than store the radioactive spent fuel, Korea would prefer to follow Japan's plan to recycle spent fuel in FBRs. Yet this would further accelerate the production of plutonium in Northeast Asia, amid the considerable insecurities that will likely pervade the region as Korea moves toward political reunification.

China too has embraced the fast-breeder concept, like Korea following Japan's lead. In late February 1995, the Chinese National Nuclear Corporation announced plans for a pilot reprocessing plant, to be completed—like Japan's own Rokkashō—around the year 2000. This facility would be capable of reprocessing one hundred tons of spent fuel annually.[21]

A full-fledged plant would be completed by the year 2015. With respect to the even longer term, China and Japan have already begun informal discussions on technical collaboration and cost-sharing in FBR research. Although China's decision to deploy FBRs entails no direct proliferation danger, given its existing nuclear status, it does sharply increase the prospective amount of civil-use plutonium accumulating in Asia beyond clear international control.

The plutonium-based nuclear fuel-cycle approach to energy sufficiency, related as it is to the issue of nuclear proliferation, often provokes international political tensions, regardless of the intentions of the nation pursuing it. The Carter administration initially opposed activation of Japan's experimental Tokaimura nuclear reprocessing plant in 1977, as the plant itself was about to commence operations. Only grudgingly did it acquiesce in the inauguration of the reprocessing program, on an interim basis. Although the Reagan administration was relatively nonjudgmental, Bush administration Defense Secretary Richard Cheney indirectly criticized Japan's nuclear initiative again in December 1991, in the course of more direct pressure on Korean efforts. "South Korea, the U.S., and many other countries have proven that reprocessing is not a necessary prerequisite to a legitimate civilian nuclear energy program,"[22] he suggested. Cheney's ostensible reference point was North Korea, which Washington has strongly pressed to abandon its reprocessing facilities. But he could just as readily lodge the same criticism against Japan, or any one of several Asian countries for which the logic of plutonium-based nuclear fuel-cycle development—the Plutonium Economy—is ever more attractive on purely economic grounds. The United States expressed similar, more sharply focused concerns about South Korea's nuclear program in November 1992.[23]

China's Bomb and Its Implications

The technology and the raw materials for nuclear weapons, in short, are more and more readily available across East Asia, as in other global hot spots like the Middle East. The key variable in determining the likelihood of dangerous proliferation is ultimately political will. How intent will the many technically competent, energy-short nations of East Asia be in coming years on pursuing the bomb?

One whose position is already established, and that sends shivers of apprehension through many of the others, is mainland China. Since exploding its first crude twenty-five-kiloton nuclear device from atop a steel tower in the isolated Gobi Desert during the Tokyo Olympics of October 1964, China's nuclear program has proceeded rapidly, even amid the confusion and hysteria of the Cultural Revolution. Just thirty-two months after its atomic explosion, China in June 1967 successfully tested a full-yield hydrogen bomb dropped from an airplane. CHIC-6, as it was code-named in Washington, yielded 3.3 megatons, or 231 times the power of the Hiroshima blast.[24]

Over the past quarter century, China has steadily upgraded both the power and precision of its nuclear weapons and the sophistication of their delivery systems. After perfecting IRBMs capable of striking assorted Soviet and East Asian targets around 1970, it successfully launched an ICBM over nine thousand kilometers into the South Pacific in May 1980, putting the United States also within range of its weapons. Thirty years after China's work on nuclear submarines began, it also succeeded at the underwater launch of long-range missiles from its Xia-class submarines in September 1988.[25] In the same year, China also successfully tested a neutron bomb, lethal against humans without affecting structures, whose deployment with the U.S. military the Carter administration had rejected on humanitarian grounds a decade earlier.[26]

China's nuclear weapons and nuclear delivery systems are continuing to develop. Alone among the nuclear powers, China continued systematic nuclear weapons testing throughout the first half of the 1990s, although its overall record of testing remained far less extensive than the traditional superpowers'. By the year 2000, China's armed forces are expected to be equipped with cruise missiles and increasingly elaborate space-based defense systems, albeit with capacities still markedly inferior to those of the United States.[27]

China's original 1964 A-bomb explosion sent shudders of anxiety across East Asia, on the very eve of large-scale U.S. intervention in Vietnam. China thus provoked all the non-Communist nations of Northeast Asia to explore nuclear weapons as an active option. Its detonation of a 3.3-megaton hydrogen bomb less than three years later intensified these concerns, prompting Japan to unobtrusively keep its options open. Despite the "nuclear allergy" of its public, Japan began quiet work on broader nuclear options in 1965, just after the first Chinese test. Despite a public declaration of "three nonnuclear principles" in 1968[28] that ultimately won Prime Minister Satō Eisaku the 1974 Nobel Peace Prize, Japan's Ministry of Foreign Affairs urged, in a secret 1969 document, that Japan maintain its nuclear-weapons-making potential.[29]

In its first national defense white paper, published in 1970, the Japanese government stressed that possession of defensive, small-size tactical nuclear weapons was not in conflict with the constitution, although it forswore them as a matter of policy. Although it formally signed the Nuclear Nonproliferation Treaty (NPT) in 1970, Japan did not actually ratify the treaty until more than six years later.[30] Time after time, NPT ratification passed the lower house, only to subsequently languish in the House of Councillors.[31]

South Korea and Taiwan, both similarly unnerved by Chinese nuclear capacity and guarded in their view of U.S. guarantees, also began to quietly explore the nuclear option. Although Taiwan ratified the NPT expeditiously, in only eighteen months, in 1969 it bought a forty-megawatt Canadian research reactor, similar to the one that India used to generate plutonium for its 1974 test.[32] Taiwan almost simultaneously went to the United States requesting a reprocessing plant, but was rebuffed. Shortly thereafter it began negotiations with the French over a hundred-ton-a-year reprocessing plant, but was again frustrated by strong Nixon administration pressure on France not to provide it. South Korea, meanwhile, explored similar options, and like Japan hesitated for many years—nearly seven—before ratifying the NPT.[33]

Northeast Asia's flirtation with the nuclear option thus has a long history, extending to every nation in the region—not just North Korea. But countries seem to wax and wane in the seriousness with which they consider this possibility. An underlying energy vulnerability and apprehension about a nuclear China seem common to all. Beyond those universals, diplomatic isolation, the prospect that other regional powers might go nuclear, and doubts about the credibility of U.S. nuclear guarantees, when these factors emerge, also appear to provoke local impulses toward proliferation.

Since these three factors tend to co-vary with one another, Northeast Asian nuclear proliferation tendencies usually occur in region-wide waves, across several nations at once. The tendency was pronounced in the late 1960s and the early 1970s; there is danger that it may recur in the coming decade if U.S. nuclear policies toward North Korea, or U.S. security guarantees in relation to China, are mishandled today. Henry Kissinger argues, "Failure to resolve the North Korean nuclear threat in a clear-cut way will sooner or later lead to the nuclear armament of Japan—regardless of assurances each side offers the other."[34] Regardless of the ultimate resolution of this issue, an ambiguous outcome clearly intensifies pressures in this direction, both in Japan and in surrounding nations.

North Korea, the "hermit kingdom," has been isolated from the broad global political economy since its very foundation, fortified by a philosophy of *juche*, or self-reliance. Since the waning of the Cold War and the collapse of the Soviet Union, the North's longtime trading partner and political supporter, this isolation has intensified. South Korea's deepening ties with both Russia and China, with which Pyongyang conducted over half its foreign trade in 1990, have made things worse. So has the stagnation of its shriveling economy, now less than one-tenth the size of South Korea's.

Despite formidable conventional military capacity, including an army of well over a million men, North Korea since the late 1980s has thus held strong and deepening incentives, rooted in national isolation, to go nuclear. Nuclear weapons will assure that it is treated seriously, as a major player in the region, even if its economy is in crisis, it lacks close allies, and the United States stands strongly behind South Korea.[35] Conversely, however, the economically feeble and diplomatically isolated North has few other cards in its hand. It will thus likely cling quite tenaciously to at least some residual nuclear capacity, as the tortuous 1993–94 negotiations with the Clinton administration and their equally frustrating aftermath clearly showed.

Diplomatic isolation between the superpowers was, of course, a factor behind China's original decision to build a bomb. It has also influenced Taiwan significantly, especially since the United States, its longtime guarantor, recognized mainland China in 1979. Deng Xiaoping, of course, firmly warned that a clear Taiwanese nuclear capacity would be, together with a declaration of independence from the mainland, one of the few conditions that would trigger Chinese military action against Taiwan.[36] But Taiwan nevertheless has strong incentives, to the extent that it is isolated, to develop ambiguous independent nuclear capabilities—similar to those of Israel, India, or North Korea, for example—as a substitute for formal international support. Its Canadian-origin Lung Tan reactor, just twenty miles southwest of Taipei, has now been operating for over twenty-five years, reportedly generating enough plutonium-bearing fuel for more than ten weapons.[37] Some reports indicate that Taiwan has also been building a secret reprocessing facility and shaping bomb cores.[38]

These covert, ambiguous actions—gradually enhancing the credibility of a Taiwanese nuclear option—may well continue, and even intensify, as the economic and military profile of mainland China rises. Taiwanese president Lee Teng-hui's July 1995 threat, amid China's aggressive missile testing, to possibly revive Taiwan's military nuclear program—a threat later repudiated—can be seen in this context. Ambiguous threats are Taiwan's only recourse in the face of two sobering alternatives. If too clear-cut and resolute, Taiwanese nuclear steps might prompt a Chinese preemptive strike. Yet clear Taiwanese inaction would also conversely breed the dangerous perception on the mainland that it could act with impunity against Taipei. Drifting in a vague, complex world of innuendo and rising technical capacities, nuclear relations across the Taiwan Strait will likely remain a dangerous element of Northeast Asia's security picture—and one decidedly impervious to outside attempts at peacemaking—for many years to come.

To Build or Not to Build?
Strategic Dilemmas for American Allies

South Korea and Japan stand in a somewhat different political-military relationship to the nuclear option from North Korea and Taiwan, although they too have intermittently taken more extensive steps toward that delicate threshold than often realized. Both nations are unambiguous military partners of the United States, linked to Washington by treaty, and both station U.S. forces on their soil. Although the same energy vulnerabilities, fears of China, and spinoffs from civilian nuclear programs shape their strategies as affect Taiwan, the nuclear policies of South Korea and Japan must inevitably turn much more heavily on the credibility of the formal American defense guarantees that they so unambiguously possess.

Many respected strategic theorists argue that Japan's own geographical configuration and its close proximity to the three major nuclear powers (the United States, Russia, and China) make an independent nuclear capacity irrational and force Japan into alliance relationships with at least one of the other neighboring nuclear powers.[39] Fundamentally, the argument is that a small, densely populated island nation the size of California is simply too vulnerable to enemy attack to be a credible nuclear power, no matter what sort of nuclear armament it might in fact possess. It would suffer extraordinary damage and loss of life should even a single high-yield enemy nuclear weapon reach its shores. Accordingly, there is a strong strategic rationale, economics aside, for such a vulnerable nation to be allied with a country like the United States, with which it shares common values and economic objectives much more fully than with alternative potential allies, such as Russia and China. If this logic holds, there should be an intrinsic value for Japan in American guarantees, if only Japan can be certain that the United States will uphold its commitments. A parallel logic should prevail for South Korea.

Yet other respected strategists disagree. They insist that whatever the strategic rationality, Japan, at least, will ultimately become a nuclear power. Often influenced by realist assumptions about nations striving ultimately to maximize their power resources in international relations,[40] such theorists doubt that Japan, as an economic superpower, will indefinitely restrain the urge to translate this economic leverage into military power with a turn at the nuclear chessboard. Henry Kissinger and Kenneth Waltz, for example, foresee an increasingly turbulent world of waning American predominance at the close of the twentieth century, in which Japanese rearmament is virtually inevitable and acquisition of nuclear weapons is not unlikely.[41]

Despite the rhetoric of summit meetings, there has been clear and troubling oscillation in the perceived credibility of U.S. commitments toward Northeast Asia over the thirty years since China got the bomb. The most delicate period so far was clearly the 1970s, particularly the latter half of the decade. Nixon's secret diplomacy with Beijing, the abrasive economic overtures of John Connally, Vietnamization, the fall of Saigon in 1975, and, subsequently reversed, Jimmy Carter's 1977 decision to withdraw U.S. troops from South Korea all damaged U.S. credibility with its Northeast Asian allies in different ways.

Shadow Nuclear Power

The result in South Korea and Japan, not surprisingly, was quiet, ambivalent explorations of the nuclear option, against the backdrop of public opinion that saw such a development as virtually inevitable, however unpleasant it might be. In October 1969, for example, *Yomiuri Shimbun* conducted a poll in which 77.8 percent of respondents predicted that Japan would possess nuclear weapons by the year 2000, with only 8 percent believing it would not.[42] Parallel research conducted during 1965–68 among Japanese college students suggested that nearly two-thirds, or 64 percent, believed that Japan would develop nuclear weapons by the mid-1980s.[43]

In late 1971 or early 1972, primarily in response to Richard Nixon's decision to withdraw the U.S. 7th Infantry Division from Korea, and also no doubt fearing the ambiguous mood in Japan, President Park Chung Hee decided to act.[44] South Korea initiated negotiations in 1972 with France to buy a reprocessing plant. When this deal became public in 1975, Seoul maintained that the technology was needed for energy security and to match Japan's Tokaimura reprocessing plant, then under construction. Despite a safeguards agreement among the IAEA, France, and South Korea, the United States intervened strongly and negatively. Henry Kissinger finally stopped the Korean program by brusquely but clearly informing President Park that the United States would cancel its security commitment to the ROK if the South persisted with its nuclear weapons program.[45] Seoul finally ratified the NPT, after equivocating for seven years.

In the late 1970s, South Korea revived its consideration of the nuclear option, after President Jimmy Carter announced U.S. intent to withdraw troops and nuclear weapons from the South by 1980. It also sought expanded ballistic missile capabilities, trying vainly to acquire the seven-thousand-kilometer-range U.S. Atlas Centaur missile in 1979.[46] The South likewise persistently sought reprocessing

technology, forcing the Reagan administration to press Canada to end cooperation with Seoul in 1983–84, and the Bush administration to intervene as well only
five years later.

Despite Japan's "nuclear allergy," as the only nation ever to have suffered
actual nuclear bombardment, its policies have subtly fluctuated with its confidence in the United States. As in the case of South Korea, the most delicate period
was the 1970s, when Japan failed for several years (1970–76) to ratify the NPT
and argued with the Carter administration over the Tokaimura reprocessing plant,
as noted above. Japanese leaders and the Japanese Ministry of Foreign Affairs
have also periodically called attention to Japan's latent nuclear capabilities, as
Prime Minister Hata Tsutomu did in mid-1994,[47] in order to maintain diplomatic
leverage through this shadow capacity.

Politics and the Nuclear Threshold

Preceding pages have suggested complex and dangerous international dynamics
on nuclear issues in Northeast Asia, flowing in part from the economic attractiveness of the plutonium-based nuclear fuel cycle. The Pandora's box could well
bring not only Japan but possibly several other Asian nations to the nuclear
threshold over the coming decade. These nuclear dynamics could interact with
rising tensions over the energy resources of the East and South China seas, where
Japan, China, Vietnam, Malaysia, the Philippines, and Taiwan, among others, all
have conflicting territorial claims aggravated by a shortage of domestic energy
reserves and rising domestic demand.

Will these dynamics predispose Japan, both Koreas, and possibly Taiwan
increasingly toward acquiring some sort of nuclear capacity? How might the dangers be amplified by the proliferation of increasingly accurate long-range missile
capacities in the region? At the very least, lingering uncertainties will compel the
nations of Northeast Asia to keep their options dangerously open, spreading increased mutual anxiety across the region.

Troubling new possibilities may be presented by the nuclear fuel cycle, unless
new multilateral arrangements evolve to neutralize them. Japan will accumulate
large amounts of plutonium simply in the process of operating civilian reactors,
especially fast breeders. Korea, China, and Taiwan could do likewise should they
proceed down the FBR route. The danger that an East Asian plutonium stockpile
might be diverted to military or terrorist use will be intrinsic, and omnipresent,
even if the key Northeast Asian nations exhibit no direct intention to go nuclear.

The dangers for the region will be compounded by political uncertainties,
centering on the complex future of Korea. In this regard, North Korea's remaining

residual nuclear capacity, unclarified despite the 1994 U.S.–North Korean agreement, remains troubling, although realistically difficult to suppress, given its importance to North Korea's overall political-economic standing in the world. How much North Korean nuclear capacity would it take to induce South Korea to arm? If Korea were reunited, would it keep this apparent Northern capacity? What would happen to Korean nuclear potential if U.S. forces withdrew, for whatever reason, in the wake of reunification? The danger of nuclear conflict or accident could cloud virtually all scenarios for the future of the troubled Korean peninsula.

In the end, the most important nuclear issue in Asia's future—one that could have global consequences as well—is the orientation of Japan. Domestic political developments, interacting with a broader Asian regional environment of rising tensions outlined here, will heavily influence that. But the issue may be configured somewhat differently from the way it has been over the fifty years since Hiroshima.

The evidence is overwhelming that the majority of the Japanese people do not presently want Japan to possess nuclear weapons, and that this feeling is considerably stronger than it was even in the early postwar years when the memories of Hiroshima and Nagasaki were still fresh. In 1955, for example, 23 percent of the Japanese public favored the acquisition of nuclear weapons—more than double the percentage of the early 1980s, according to NHK Broadcasting.[48] Only a tenth of Japanese believe that nuclear weapons can be abolished from the earth, yet they themselves are reluctant to possess them.[49] A solid majority supports Japan's so-called three nonnuclear principles, declared in 1968 by Prime Minister Satō Eisaku, amid the Vietnam War and the discreet beginnings of Japan's nuclear fuel-cycle program. Those principles, instrumental in winning Satō the 1975 Nobel Peace Prize, stipulate that Japan will not make or deploy nuclear weapons, nor will it allow them to pass through its territory.

The wartime generation that actually experienced Hiroshima, and the occupation generation that was indelibly impressed through early postwar education with the evils of wartime militarism, however, will gradually pass from the scene over the coming decade. As these changes silently occur, what impact might they have on Japan's longtime "nuclear allergy"? Will it be transferred to succeeding generations? Or will the nuclear allergy be gradually transformed into a prudential judgment of what is best in geopolitical terms for Japan, at any given point in time? As the years pass, Japanese diplomats may grow more prone to tacitly manipulate their country's status as a "shadow nuclear power" for national advantage, and less inhibited about cautiously developing nuclear and related strategic options, should international circumstances dictate. And the Japanese public will likely be more explicitly critical of the nuclear policies of other major nations.

Japan is much larger economically, and more sophisticated technologically in most respects, than four of the five current declared nuclear powers, including Britain, France, Russia, and China. It sees no special legitimacy in an exclusive nuclear "club" that excludes it, and resents efforts by that club's members to use their possession of the bomb as a rationale for special consideration at the United Nations and elsewhere. Unless Japan sees clear efforts toward reducing the role of nuclear weapons in international affairs, it may be increasingly prone to either manipulate its shadow nuclear standing diplomatically or, in an extremity, go nuclear.

These emerging patterns in Japanese thinking are already apparent, albeit more in elite than in mass opinion. Japan, for example, showed considerable ambivalence at the 1993 Tokyo summit about the indefinite extension of the Nuclear Nonproliferation Treaty, citing a range of concerns centering on the North Korean nuclear program. At the forty-sixth anniversary commemoration of the Hiroshima nuclear attack, on August 6, 1993, Hiroshima mayor Hiraoka Takashi, a Socialist, actually exhorted his country's leaders to reject an indefinite extension of the NPT, pressing for a more radical attack on nuclear weapons themselves, rather than only on their acquisition by new nuclear powers.[50] Right-wing members of the conservative Liberal Democratic Party, such as Ishihara Shintarō, author of the best-seller *The Japan That Can Say No*, likewise argue for nuclear weapons on nationalist grounds, as the accepted trappings of great-power status, and as a repudiation of abject subordination to the United States.

Pragmatists, leftists, and rightists in Japan thus all converged during the early 1990s in an attack on the NPT, about which Japan had also shown significant ambivalence in the 1970s, despite its "nuclear allergy." Japan did, to be sure, ultimately vote in the spring of 1995 to indefinitely extend the NPT. Yet lingering frustration with the disjuncture between its strong economic and technological capabilities, on the one hand, and its lack of influence on the nuclear chessboard still remains in some diplomatic and political circles. Such frustration was compounded by the extended French and Chinese nuclear tests of 1995–96, announced insensitively close to the fiftieth anniversary of the Hiroshima nuclear bombing.

In considering the fateful nuclear question it must at least be noted, however, that the general Japanese public itself appears not to expect nuclear rearmament in the near or intermediate future, despite the diplomatic and political maneuverings of its leaders. Indeed, an April 1995 *Nikkei* public opinion poll indicated that only 11 percent of Japanese respondents expected that Japan would have nuclear weapons within a decade, as opposed to 66 percent of Americans.[51] This ratio of the Japanese public anticipating a nuclear future is sharply lower than the analogous proportions during the late 1960s, when as noted earlier over three-quarters of respondents in some polls thought Japan would go nuclear by the year 2000.

The Issue of Delivery Systems

The prospective strategic consequences of Japanese decisions in the nuclear area are being magnified by its growing achievements with respect to aerospace. As an island nation whose principal current security challenges come from the air, Japan has begun to develop highly sophisticated antiaircraft and antiballistic missile defense systems. Its BADGE over-the-horizon radar, deployed in the early 1990s by Nippon Electric, is one of the most advanced in the world, relying on Japan's strength at the computer-communications interface, and is leveraged by the accuracy of Japanese missiles. It is conspicuously superior to China's relatively primitive air defense system.

Japanese Tomahawk missiles, although produced under license from the United States, are said to be more accurate even than the American originals launched against Baghdad in the Gulf War. Japan's Patriot and Aegis antimissile systems are also highly advanced,[52] and it has begun deployment of AWACS airborne control aircraft. Japanese firms such as Ishikawajima Harima (IHI) participated actively in the Star Wars space-defense program of the Reagan administration, absorbing technology that should be of continued value to Japanese antimissile defense efforts even as the United States deemphasizes such priorities. Antiballistic missile (ABM) defenses, which magnify any residual nuclear capabilities Japan may develop, have a particular political attraction in Japan in that they are overtly defensive in character and hence clearly within the purviews of Japan's current "no-war" constitution.[53]

Japanese antimissile defense efforts were given additional stimulus by North Korea's launch of an intermediate-range ballistic missile in May 1993. This new missile, the Rodong 1, flew five hundred kilometers (310 miles) from its base on North Korea's southeast coast into the middle of the Sea of Japan. But as was widely repeated in the Japanese press, it reportedly has a range of one thousand to thirteen hundred kilometers (620 to 806 miles), giving it the capacity to hit Japan's second-largest city of Osaka.[54]

In the wake of the test-firing of North Korea's Rodong 1, the Japanese government announced that it was studying the feasibility of introducing a limited

"Theater Missile Defense System" in its next five-year defense plan, beginning in fiscal 1996.[55] Such a system would involve placing sensors in space or on Airborne Warning and Control System (AWACS) aircraft to detect incoming missiles early enough to launch countermeasures, such as a Patriot missile. Japan has, as indicated above, six Patriot missile units already, and plans to deploy two AWACS planes in fiscal 1997. The Theater Missile Defense System was an aspect of the U.S. Strategic Defense Initiative (SDI) on which U.S. and Japanese firms had worked jointly that survived the Clinton administration's May 1993 cutbacks, and could be an initial element of a subsequent Japanese antiballistic missile (ABM) system. Its realization, however, may well be delayed by objections from China, political cross-currents within Japan, issues of technology transfer with the United States, and the high costs of both development and implementation.

If Japan were ever to develop global strategic capabilities, two critical elements of a diversified delivery system would be long-range land-based booster rockets and nuclear submarines. Neither is really required under the current, relatively limited formal conceptions of Japanese defense, as symbolized in the 1991–96 midterm defense plan. Neither is being vigorously promoted at present. Yet both are being slowly developed, under civilian-oriented rationales. And both have vigorous proponents within the Japanese uniformed military.

Japan's National Space Development Agency (NASDA) has spent over $2 billion on development of the 260-ton H-2, the first rocket made exclusively with Japanese technology. After launching twenty-four rockets without a failure with American assistance, NASDA encountered problems during 1992–93 in independent efforts at using the H-2 to launch large-scale satellites.[56] Yet in early February 1994 the 165-foot, two-stage rocket, developed by a consortium of seventy Japanese companies without any foreign assistance whatsoever, blasted smoothly into orbit, just after dawn, from Tanegashima Space Center in Kyushu. With a new high-tech engine similar to that of the U.S. space shuttle and special composite materials in the fuselage, the H-2 is capable of launching two-ton payloads into orbit.[57] It could easily be the forerunner of a Japanese ICBM, should Tokyo decide to build one, since all the technology is Japanese, and Washington has no authority to slow development by withholding licenses.

Development and production costs for the H-2, as well as launching expenses (nearly double those of U.S. and European producers), are so high that this booster, and the Japanese rocket-launching program more generally, are hard to justify on commercial grounds.[58] Yet Japan nevertheless is proceeding further. The February 1994 H-2 launch carried an orbital reentry experiment (OREX) payload intended to help design the next element in Tokyo's space program: a Japanese version of the U.S. space shuttle. Despite its economic inefficiency, Japan's space effort continues vigorously forward, fueled by national pride, interest in technological spinoffs, and hedging against declining U.S. interest in manned space exploration. It also keeps Tokyo's geostrategic options open.

Japan's conventional submarine fleet, whose oldest vessel was commissioned in 1976, is considered one of the most modern in the world, and there are technical efforts under way to develop its nuclear capabilities as well.[59] Research is currently under way on two nuclear reactors with naval defense implications—one for icebreakers and another for deep-sea research submarines. The Japanese navy, or Maritime Self-Defense Force, as it is formally known, already operates one large icebreaker, the *Shirase*, for which the nuclear reactor under development may be especially suitable.

Japanese submarines, although modern, are at present not configured for nuclear power, and prevailing antisubmarine warfare strategy of waiting to ambush enemy Russian subs at the three vital choke points where they could emerge from Vladivostok to the open Pacific does not particularly require them. But when strategy changes to protect Japan's broader sea lines of communication, which could well occur in the context of maritime rivalries with China, long-distance capabilities could become increasingly attractive, and nuclear submarines are uniquely well suited for such a role.

In the final analysis, the constraints on Japanese nuclear capabilities, both with respect to electric power and defense applications, are political rather than technical. Japan could easily have a military nuclear capacity, and ultimately the delivery systems to make it internationally credible. It is restraining itself politically. How long will this self-imposed restraint last, and under what circumstances might it end?

JAPAN'S STRUGGLE FOR STRATEGY

If one wants psychological distance from the world of military and geostrategic affairs, there are few better places to find it than strolling the streets of Tokyo, the economic heart of Asia for three generations and more. One can start from the Imperial Palace, so fatefully spared devastation in World War II, and wind one's way through the heart of the city, past MacArthur's old headquarters in the Dai Ichi Life Insurance Building, westward toward the government offices of Kasumigaseki. Approaching the low hills of Moto Akasaka, with its massive, looming American embassy building, erected on the site of an old imperial military headquarters, one passes a cluster of economic and political satrapies that surround it. There is the Japan Export Trade Association, MITI's Bicycle Promotion Association, the headquarters of the Kochikai, former prime minister Miyazawa Kiichi's political faction, and the Hotel Okura, temporary home to Bill and Hillary Clinton at the 1993 Tokyo summit. To be sure, one also passes en route great expanses of uniformly new, unremarkable buildings, obscuring sites where fifty years before there were only smoking remnants left by General Curtis LeMay's incendiary B-29 attacks. But history has mercifully buried those reminders of a turbulent and painful Japanese past.

In the course of a stroll some blocks farther through Roppongi, one of Tokyo's prime nightlife areas, one comes upon Japan's Pentagon: a ramshackle collection of dilapidated, low-lying buildings—many dating from the 1930s and seemingly untouched since. The nighttime tranquillity of this unlikely military complex contrasts sharply not only to the mah-jongg parlors and French restaurants across the street, but also to the bustle of the Ministry of International Trade and Industry just a few blocks away. There lights blaze deep into the night as economic strategy for the future is plotted.

Apart from the Japan Defense Agency home offices and a single operations command at Ichigaya inherited from World War II, Tokyo is virtually devoid of military presence. Casual visitors—indeed, even natives—can go for weeks, or possibly years, without passing military personnel on the streets in uniform. They will hear "Kimigayo," the Japanese national anthem, only at sumo matches, or late at night on television, as NHK, the national broadcasting network, ends its day.

This same pacific pattern pervades virtually all of the Japanese archipelago. The uniformed military, just like strategic issues, are simply submerged, both visually and in the mass intellectual life of the nation. Japan, to all appearances, is as peaceful and divorced from geostrategic concerns as the New Hebrides, New Caledonia, or any other South Pacific island backwater.

Japan's Rising Military-Industrial Potential

Yet quietly, over the past three decades, Japan has forged the sinews of defense-industrial strength that would easily provide it with potent military capability should it decide to take that course. Its GNP has risen from under 1 percent of the global total in 1955 to just over 3 percent in 1970 to 15 percent today: one-sixth of the entire economic output on earth. Its overall industrial production is double that of one fallen superpower (the former Soviet Union) and could well surpass that of the United States within a decade.

To be sure, Japan has consciously subordinated defense production to other goals. Total weapons procurement in 1994 came to less than 0.5 percent of total industrial production, an amount equaled by sales at the nation's sushi shops.[1] The national ice cream and pinball machine industries are also roughly as big.[2] No major firm relies on defense procurement for more than 25 percent of total sales.

Over time, however, Japan has relentlessly developed a broad range of high-technology sectors closely linked to battlefield capabilities in the post–Cold War era now dawning. Some of the most strategic lie in the arcane, high-growth field of optoelectronics, where Japan's traditional strengths in optics and electronics converge. In the civilian sector these optoelectronic strengths have given birth to world-class camera, robotics, television, copier, and integrated-circuit industries, with major emerging spinoffs for autos as well.

Yet many of the same optoelectronic sensors and guidance systems that produce high-quality 35mm photos can, in high-performance incarnations, also guide a supersonic missile unerringly to its target. It was, after all, a smart bomb armed with a Sony minicamera and major amounts of Japanese guidance electronics that destroyed the Red River bridge in Hanoi during the Vietnam War, after F-4 Phantom pilots with purely human skills had failed for five years to do so. Japanese optoelectronic components were also at the heart of American Tomahawk missiles that rained on Baghdad during the Gulf War.

Japanese optoelectronics in its military incarnation—precise, sophisticated sensors and guidance systems for missiles, aircraft, and early-warning devices—will likely be ever more important on the battlefields of the future, with their emphasis on mobile, often long-distance conflict on the sea, in the air, or even in space. Other Japanese strengths in miniaturization, automation, telecommunications, and the development of durable, lightweight advanced materials will also leverage these optoelectronics capabilities and enhance their potential military importance. So will Japan's strong record in quality manufacturing. In all of these areas, Japan ranks first or second in the world.

The strategic importance of Japan's latent military-industrial capabilities can well be seen in their ability to shape the production cost and performance of supersonic jet fighters, even though Japan still does not autonomously produce the fighters themselves. More than half the production cost of a top-of-the-line F-18 jet fighter, for example, already goes for advanced electronics and communications equipment, most of which Japanese manufacturers can produce to world-class standards. The F-18's combat potential is also enhanced greatly by its light weight and extraordinary maneuverability, including the ability to withstand heavy G-forces while twisting and turning in a dogfight. Carbon-fiber wing reinforcement, titanium engine components, and composite-material body construction are all cutting-edge technologies, still evolving, in which Japanese industrial sophistication continues to have major defense implications.

On the battlefields of the future, more and more military equipment—on land, in the air, and on the sea—will be unmanned, as combatants go to ever greater lengths to minimize casualties.[3] The more pronounced this tendency, the more important Japanese strengths in automation, miniaturization, and optoelectronics will become to weapons designers and military tacticians worldwide. For the essence of Japan's civilian technological drive—driven by high labor costs and high economic growth in an overcrowded land—has long been to reduce the use of human labor. Intelligent, yet dispensable and low-cost robot warriors may well first materialize in Japan, or at least through cooperation with Japan's defense-industrial base.

Over the past decade, Japanese industrial strategists, in both the public and private sector, have stressed the importance of the "C and C" interface between computers and communications.[4] In this respect, they have placed particular emphasis on the development of commercial satellites, optical-fiber data transmission, and mobile communications. All have increasingly important military implications.

As the concern of defense planners moves toward various space-related contingencies, accurate satellite sensing and communication grow more important. As land-based data-transmission requirements also rise in the increasingly automated warfare of the future, optical-fiber transmission lines that protect military communications from many electronic countermeasures or eavesdropping will likewise grow in importance. Japanese firms hold roughly half the global market in this strategic infant industry, producing the purest optical fiber, with the highest-performance communications potential, in the world.

Japan's capabilities in dual-use technology and dual-use production in electronics, telecommunications, precision machinery, and basic materials give it dual military-industrial capabilities: both to develop a potent arms industry at home and to feed the military machines of others.[5] Given Japan's economic power, it could also easily support a substantially greater military establishment than at present. Japan now spends over $50 billion annually on defense—more than total U.S. military expenditures in the Pacific, and more than half of the entire Russian defense budget. Yet this huge sum is still less than half the share of GNP that the United States devotes to its own military.[6]

Japan thus has enormous latent potential both to expand defense spending and to aid the efforts of others, including potentially China, Russia, or Europe as well as the United States, should domestic political currents so dictate. Indeed, Japan's future strategic choices will have fateful regional and global consequences. What will it decide to do?

In thinking about future Japanese defense choices, it is important to consider the long-term forces that skew and predetermine the decisions that Japan consciously does make. Two factors are clearly Japan's rising economic and technological strength, alluded to above. They will make autonomous military capabilities easier than in the past, should the conscious choice to develop them be made. And the very ability to easily and rapidly produce state-of-the-art missiles, warships, and even nuclear weapons could in turn encourage their development, through a process of techno-political "creep."

Another important consideration is the "technology and security ideology" with which Japanese approach their emerging national security requirements.[7] Reflecting Japan's standing as a late developer that was thrust into the international system to sink or swim near the high noon of nineteenth-century Western imperialism, the Japanese bureaucracy has come to see autonomous technical and industrial capacities as central to national security. Much of the rest of the nation clearly believes this as well. The result is a historically rooted, yet now almost instinctive impulse toward autarky in technical matters. In defense, this means a bias toward independent development of weapons systems, or domestic licensed production when that cannot be achieved. This bias persists even when such a stance is clearly not cost-effective or politically expedient in international diplomatic terms.[8]

This autarkic tendency was starkly clear in the early stages of the 1989 U.S.–Japan controversy over the building of the next-generation FSX fighter.[9] Until the Reagan-Bush administration, backed by the U.S. Congress, exerted strong lobbying pressure against the concept of autonomous development, MITI, the Japan Defense Agency, and the Japanese defense industry fully intended to produce their next-generation fighter independently of the United States. Only after a bitter struggle and the forceful intervention of Prime Minister Takeshita Noboru did they back down.

Even after the formal decision on codevelopment was made—under terms heavily influenced by the U.S. Congress—latent resistance to U.S.–Japan codevelopment continued at the working level in Japan. The FSX project fell more than two years behind schedule, with even its wooden prototype not being unveiled until July 1992, and the mechanically complete version in early 1995.[10] The finished aircraft are unlikely to be deployed until close to the year 2000, at a projected cost that could surpass $100 million each, or four times the cost of the American F-16 fighters on which the FSX is based. By early 1995, as the first finished prototypes reached completion, the FSX project had already cost nearly $3.3 billion, or more than twice the amount originally contemplated. And the United States and Japan had still not agreed on two critical points: the number of planes to be produced, and which of eight particular Japanese technologies used in the FSX should be available to the United States. The bitter technonationalistic struggles of six years earlier were still not clearly resolved.

Japanese impulses toward independent military development have progressed even farther in the missile sector than they have in aircraft. Japan, which now has multiple domestic missile producers, including auto manufacturer Nissan, has developed replacements for U.S.–supplied Stingers (the hand-held missiles so effective in Afghanistan), Sidewinders (air-to-air interceptors), and

Harpoons (antisubmarine missiles).[11] Its Tomahawk and Patriot missiles, still produced under license in Japan, are reputedly more accurate than even those used against Baghdad during Desert Storm. Japanese autarky in missile technology was given a big boost in early 1994 with the successful launch of the new, totally domestic H-II booster rocket, which could also potentially serve as an ICBM.

A moment of truth in the U.S.–Japan defense procurement relationship now looms, in the coming decade. Around the year 2005, Japan will face a procurement decision on a new generation of surface-to-air (SAM) missiles, to replace the U.S.–designed Patriot. Shortly thereafter it must begin thinking beyond the bitterly contested FSX fighter. There is also the impending question of theater missile defense (TMD). How will the ongoing tension between international political relationships and the century-long drive for technological autonomy resolve itself in a world where Japanese technical strengths and economic prowess will be even greater than they are now? The answers to such questions, and to how Japan will ultimately deploy its huge but thus far largely latent defense potential, clearly lie in the "struggle for strategy" now quietly deepening in Tokyo.

The Deepening Debate over Japan's Future

Below the placid surface, a tortured debate rages across Japan regarding its future. On substance, Japanese are deeply divided, with those favoring and those opposing an expanded military and more proactive foreign policy delicately balanced, amid numerous, complex cross-currents. But on one point there is virtual consensus: status-quo institutions and diplomatic processes are inadequate to handle the deepening long-term international problems confronting Japan today.

This frustration and intellectual ferment seems light-years from the first Tokyo summit of 1979, when Japanese stood in awe and self-congratulation at the appearance of the G-7 leaders in their midst and lapped up 700,000 copies of Ezra Vogel's *Japan as Number One*[12] in a matter of months. The present even seems distant from the smug self-confidence of the second Tokyo summit (1986), when Japanese basked vicariously in the warmth of their prime minister's easy "Ron-Yasu" relationship with Ronald Reagan, and in the prosperity induced by an expanding bubble economy and buoyant exports to America.

The intellectual challenge to Japan's traditional transpacific moorings, and the urgent sense that something new is needed in Japan's relations with the world, is only a decade old. It has its early origins in the reencounter with Asia of the late 1980s, as a doubling in the yen's value during 1985–87 pushed Japanese investment and Japanese travelers out sharply into a region they had begun to regard again as foreign, after the intimacy of the wartime years. To be sure, there had been the shock of recognition and fellowship with nearby nations in the late 1970s, as China began to open and as Japanese orphaned by the Russian Red Army in China at the end of World War II came back to find their roots in Tokyo, Yokohama, and Niigata. But Japan's serious reencounter with Asia really began in 1987 and thereafter.

A second thread was domestic frustration: simultaneous disquiet at Japanese diplomatic inaction and, to some, lack of international ideals, etched clearly in a series of crisis-management failures. Most searing was the Gulf crisis of 1990–91, when Japan failed either to commit seriously to the Allied cause—even in humanitarian terms—or to come up with a clear alternative position. Despite Japan's massive financial commitment to Desert Storm—$13 billion, or 20 percent of the overall cost, which required a tax increase in Japan itself—few foreigners appreciated Japan's contribution, a reality that frustrated the Japanese deeply. Few wanted to see their country as simply an unappreciated cash register for the United States.

A third key element in the emerging debate over Japan's global role was deepening political ferment at home. From the Lockheed scandal, which embroiled conservative ruling party strongman Tanaka Kakuei from his arrest on bribery charges in 1976 through his conviction seven years later, until the collapse of the ruling Liberal Democratic Party in 1993, the party's rule was a deepening tale of corruption and special favors, exposed ever more zealously by a powerful mass media. As big business became increasingly affluent, international, and disengaged from the precarious, debt-based "bicycle economy," it also grew more critical of the ruling conservatives. When the powerful LDP renegades Ozawa Ichirō and Hata Tsutomu broke from the ruling party in June 1993, together with several dozen followers, bringing down the Miyazawa cabinet, big business did not strenuously object. Indeed, it looked forward, like most of the mass media, to the birth of a new political order.

Electoral reforms passed in January 1994 assured that such a new system, more oriented toward mass public interests than the existing multimember districts, would ultimately emerge. To be sure, the birth pangs were sharp ones. They made Japanese politics—with three revolving-door prime ministers inside a year after the LDP's collapse—initially more turbulent than even Italy's. But many of the old structural constraints that hamstrung Japanese diplomacy were about to be released. Historic change was in the air.

What are the key forces shaping emerging Japan's worldview, as it confronts a turbulent Asia to which it is ever more tightly linked? Fundamentally, there are two elements. There is the backdrop for policy created by prevailing public opinion. This by no means determines policy, and tends to be a conservative, reactive element, shaped heavily by past traditions and media interpretation of recent events. Its major impact is to constrain policy innovation, and hence to arrest emergence of a more proactive Japanese global role. Then there is a second, more directly decisive determinant of what Japan actually does in the world: the orientations of elites themselves.

The Japanese Public: Ripe for Change?

Japanese public opinion on security matters is in many ways profoundly biased toward the Cold War status quo, contradictory though that may often be. Substantial majorities in all recent major national opinion polls, for example, believe that Japan should maintain its Self-Defense Forces at roughly current budgetary levels, and at roughly the current size. But they also believe, perhaps inconsistently, that their 1947 constitution's famous Article 9, which states that "warmaking capabilities shall never be maintained," has likewise contributed to peace.[13]

Further ratifying the status quo and its contradictions, large majorities of the Japanese public believe that the U.S.–Japan Mutual Security Treaty contributes to peace. They also support their country's three nonnuclear principles, as noted in Chapter 4, and believe that Japan's nonnuclear status will continue to be sustained in future. Although recent opinion polls suggest that a majority of Americans believe Japan will ultimately go nuclear, little more than a tenth of Japanese seem to share that view, as noted earlier.[14]

Japanese also hold a broadly favorable image of the United States. That image was remarkably stable over the course of the 1980s and early 1990s, even in the face of Japan's rising economic power and persistent U.S.–Japan trade disputes. For example, 42.6 percent of Japanese polled by the Jiji Press in January 1995 said they "liked" the United States, compared to 45.2 percent ten years previously and 37.9 percent in July 1981. More than half, in even recent polls, think the strong American cultural influence on Japan has been basically benign. Conversely, only 0.6, 10.0, and 0.1 percent respectively "liked" the longtime Cold War adversaries Russia, China, and North Korea in 1995, even less than in 1985.[15]

Despite important stabilizing elements in Japan's worldview, from a U.S.–Japan perspective, Japanese opinion polls suggest some seeds of trouble for the future. This is especially true if one assumes—as this volume strongly maintains—that the problems of the future will be very different from those of the past and may well require greater activism of Japan. The danger signals are fivefold.

First of all, there is the parochialism of the Japanese public regarding foreign affairs. Underlying that public's status quo orientation is a profound lack of security consciousness and concern, in the conventional Western sense of those concepts. Even though the Japanese public generally dislikes Russia, North Korea, and to a lesser degree China, it feels little immediate sense of threat—from them or anyone else. The sharpest difference in Japanese and Western threat perception has to do with global concerns beyond East Asia, particularly contingencies in the Middle East. In a major April 1995 *Nihon Keizai Shimbun* poll, for example, 57 percent of Americans saw the Middle East as a threat to world peace, but only 20 percent of Japanese did so.[16] Many more Americans than Japanese also regarded developments in Africa in that fashion, although the numbers were small (4.3 and 2.6 percent).

Japanese even appear to take remarkably detached, timeless views of geopolitical developments on their very doorstep. A striking case in point is the Korean peninsula. In a mid-1995 joint Japanese–Korean poll, for example, only 37 percent of Japanese respondents (as opposed to 68 percent of Koreans) thought Korea would *ever* be unified. And just 23 percent of Japanese (as opposed to 53 percent of Koreans) thought such a development would occur within twenty years.[17] Given the sharp recent deterioration of North Korean economic fortunes, most analysts outside Japan would likely side with the Korean assessment.

When asked the primary purpose of their Self-Defense Forces, around 60 percent of Japanese mention one of three domestic contingencies—(1) aiding in disasters, such as earthquakes and floods; (2) rendering human assistance, as in first aid for the elderly; and (3) quelling prospective domestic disturbances. This view was strongly, and perhaps justifiably, reinforced by the traumatic Kobe earthquake of 1995, and by growing fears of an even greater disaster in the Tokyo area. Only 46 percent, in a recent government poll, favored resisting with force, even if their country was invaded.[18] Less than 10 percent of all Japanese think that their men in uniform should serve as an army with war-making powers, a concept that would be conventional wisdom almost anywhere else.[19]

In addition, the Japanese public's sense of intimacy with America and affection for it have begun to decline, albeit from initially high levels. Conversely, active dislike of the United States has been rising. By November 1991, for example, 32 percent of Japanese, in an *Asahi Shimbun* poll, said they did not feel any intimacy with the United States, a figure up sharply from previous results.[20] A combination of backlash from the Gulf War, declining economic interdependence, and bilateral trade frictions seems to be at fault. After some modest improvement in 1993–94, these figures worsened again sharply in 1995, amid bitter automobile trade talks accusations of CIA economic spying, and an explosive rape case in Okinawa.

As their sense of identification with America declines, Japanese are also growing more skeptical of, and even hostile to, U.S. security guarantees and commitments. In a major 1995 Nikkei–Dow Jones poll, 38 percent of Japanese indicated a lack of confidence that the United States would come to their aid in a military emergency, compared to only 16 percent of Americans.[21] To be sure, 49 percent of Japanese (and 77 percent of Americans) still felt confidence in the U.S. security commitment. But following the Okinawa rape case, more than 40 percent of respondents throughout Japan in an October 12 Nikkei poll favored abolition of the U.S.–Japan Security Treaty, up from 28 percent only two months earlier. U.S. credibility with the Japanese public has clearly been waning.

Thirdly, and perhaps most ominously, a large and rapidly growing share of the Japanese public feels that U.S. forces in Japan, the linchpin of the current transpacific security structure, should be either reduced or totally withdrawn. This view no doubt follows from the growing skepticism of U.S. credibility noted above, together with concrete misgivings over cost, crime, and other irritants. In November 1989, according to a *Yomiuri Shimbun* poll, 39.1 percent of the public felt that U.S. troops should be either reduced or withdrawn, with most favoring a gradual phased withdrawal. By September 1992, in the wake of the Gulf War and intensified U.S.–Japan trade frictions, this ratio had risen to 61.8 percent,[22] and by December 1993 to 63 percent.[23] Disagreements over funding for these troops, in a post–Cold War era of rising social security costs, waning threat perception, and deepening fiscal stringency in both nations, could make this issue of American bases in Japan even more controversial with the Japanese public in future. The frustrations of sonic booms, racial incidents, crime, and slow-moving military traffic will not help much either.

The strongest dissatisfaction with the presence of U.S. bases in Japan is concentrated precisely where 75 percent of them are located, and where their presence is most strategically vital: Okinawa. Lying astride the major shipping lanes from Japan's home islands to Singapore and beyond, less than five hundred miles from Shanghai and Taiwan, and just slightly farther from Hanoi and the DMZ, dotted with major airfields and military communications facilities, Okinawa is the linchpin of America's military presence in East Asia. Yet the manifestations of that presence are distinctly unpopular with local citizens. Only 7.8 percent of respondents in an early 1995 *Mainichi Shimbun* poll of Okinawa residents viewed U.S. forces in Okinawa as "necessary," while 31 percent grudgingly considered them "unavoidable"; 24.9 percent thought them "unnecessary," and 29.4 percent "dangerous,"[24] even before the sensational September 1995 rape incident and ensuing mass demonstrations, the largest in Okinawan history, which involved on one climactic day 85,000 people, or nearly 8 percent of the entire local population. By November 1995, 80 percent of Okinawans reportedly backed total

U.S. withdrawl. The governor of Okinawa since 1990, Ohta Masahide, a historian of Okinawan World War II experiences who himself experienced the U.S. invasion of 1945, adamantly opposes the continued presence of U.S. bases, which he regards as likely to entangle Okinawa unnecessarily in a prospective future regional conflict. His refusal to sign extensions on several base leases in late 1995 provoked a serious political crisis for the Murayama cabinet in Tokyo.

The Okinawan people seem to feel very differently about the presence of Japanese Self-Defense Forces on their soil than they do about U.S. troops, despite reportedly strong and persisting resentment of how Imperial Army troops sacrificed Okinawans needlessly during battles half a century ago. In the 1995 *Mainichi* poll, 68 percent of Okinawans saw an SDF presence as "necessary" or "unavoidable," whereas only 39 percent viewed a U.S. presence that way. Yet the substitution of Japanese for American military power in Okinawa that the Okinawan people themselves would seem to prefer is a Pandora's box. It could sharply change the geopolitical dynamics of the entire East Asian region, possibly bringing China and Japan, less than thirty minutes apart by fighter plane in this region, into more direct and intense contention than is currently the case.[25]

A fourth complicating factor for the U.S.–Japan security relationship is the Japanese public's ambivalence about the formula for supporting the existing, increasingly unpopular U.S. forces already in Japan. In May 1990, as the formula for sharply increasing Japanese contributions for U.S. troop costs up to a 50-50 burden-sharing formula was being debated, 71 percent of respondents in a major *Asahi Shimbun* defense poll opposed the notion of "significantly" expanding Japanese coverage of U.S. base costs, with only 16 percent supporting the concept.[26] Forty percent in the same poll felt that Japan should construct its own defense framework, as opposed to 31 percent who advocated continued reliance on U.S. support. These sentiments grew stronger across the mid-1990s.

A fifth danger signal is the diffuse Japanese public sense of dissatisfaction and frustration at the nation's handling of recent international crises. This was especially true during and just after the Gulf War, when nearly two thirds many Japanese were negative toward government policies as supported them.[27] Japan's public rejected U.S. congressional criticism that Japan should send men and risk shedding blood, citing constitutional constraints. But it could not agree clearly on alternatives. The largest share of Japanese saw the main lesson of the Gulf War as the need for an "independent diplomacy," without specifying clearly what that might entail.[28]

The Strategic Views of Japanese Leadership

Against this backdrop of a frustrated and skeptical Japanese public, increasingly critical of U.S. forces in Japan yet unclear about other security options, how do their leaders view prospects for the future? Fundamentally, there are four geo-strategic orientations. Most status-quo-oriented are the Traders, who put economic considerations first. Allied with them in opposition to a major expansion of Japanese military capacities are the Progressives. More revisionist, both in domestic and international political terms, are the Realists and the Gaullists, whose operational policy differences currently center on their approach to the U.S.–Japan relationship.[29]

In defending Japan's longstanding commercial orientation in the postwar world, the Traders point to the stunning economic success of this strategy. As early postwar prime minister Yoshida Shigeru (1946–47 and 1949–54) prophesied, it has allowed Japan to focus on economic reconstruction and rapid industrial development, avoiding the geopolitical struggles that might otherwise have complicated access to foreign markets. Japan's low, nonthreatening political profile, the Traders contend, has helped persuade key trading partners such as the United States to accept large, persistent Japanese trade surpluses without retaliation. It has also caused them to countenance heavy Japanese inroads into high-growth, high-value-added markets for electronic components, telecommunications equipment, and precision machinery, even when those sectors relate closely to national defense. As might be expected, advocates of the Traders' perspective have been concentrated in key Japanese government economic ministries such as MITI and large, internationally oriented private firms, although some have also been prominent academics.[30]

The Progressives, mainly based in the mass media, academia, foundations, and the political world, are likewise leery of major changes in Japan's basically passive international security orientation. They emphasize the importance of sharply controlling military expansion.[31] Many are also antagonistic to the U.S.–Japan Security Treaty and the continued presence of U.S. military forces in Japan, although such positions are by no means universal, especially following the Japan Social Democratic Party's explicit mid-1994 recognition of the treaty.

Typical are the views of Takemura Masayoshi, leader of the Sakigake (Harbinger) Party and finance minister in the 1994–95 coalition cabinet of Socialist Murayama Tomiichi. Takemura argues that Japan should avoid the turbulent mainstream of international power politics, setting as its goal to become a "small shining country" that is a model for others in its peaceful, idealistic orientation, a bit like an Asian Switzerland.[32]

Unlike Ozawa, his bitter political rival, Takemura feels that Japan should sharply limit military spending and offshore deployment of Japanese forces, even under UN auspices. To defuse international pressures for entangling overseas involvement, Takemura is also reluctant to see Japan become a permanent member of the UN Security Council.

Consistent with his quietistic prescriptions for foreign policy, Takemura sees no need for developing more dynamic political institutions at home. Political reform, he insists, is a matter of restraining corruption and nurturing a more pluralistic democracy, rather than centralizing political power or creating more powerful political institutions. His ideal is a world of multiple parties and shifting coalitions hospitable to diverse views, rather than the clear-cut system of two large competing parties favored by Ozawa.

Typical of a somewhat more activist strain in the "dovish" camp is Funabashi Yōichi, influential senior journalist with the *Asahi Shimbun*. Funabashi, with wide experience in both Beijing and Washington, suggests the objective of Japan as a "global civilian power."[33] This sort of Japan would be activist on environmental, economic-development, and global macroeconomic issues, but would refrain from an expanded military role.

Like it or not, Japan, with its 15 percent of global GNP, is already a pillar of the world economy, Funabashi maintains. An isolationist or passive stance on economic matters would thus threaten global instability, also imperiling Japan itself, much as U.S. passivity early in the Depression of the 1930s harmed both America and the world economy. But an expanded military capacity, he argues, is not only unnecessary for Japan in the post–Cold War world but potentially counterproductive. The advanced industrial world has countenanced Japan's rapid growth and huge trade surpluses precisely because they pose no national security threat, Funabashi argues. That benign acceptance would be needlessly jeopardized by rearmament or assertiveness on conventional security issues, he maintains.

The Progressive conception of Japan's appropriate international role is sharply colored by a traumatic view of the Gulf War. Progressives in general see that conflict as a grotesque, unnecessary slaughter and Japan's heavy $13 billion financial support for Allied Forces in the Gulf as a colossal, tragic mistake. Had Japan previously developed a more clear-cut, proactive peace diplomacy, they argue, it could never have been pressed by the United States into such a waste of resources. The Progressives also strongly opposed Japan's postwar dispatch of Self-Defense Forces to do minesweeping under UN auspices in the Gulf, as well as subsequent peacekeeping missions in Cambodia and Mozambique. Such actions, they suggest, create dangerous precedents for a future where Japanese geostrategic intentions, as in the 1930s, might be less benign.

Two articulate groups of Japanese—the Realists and the Gaullists—propound much more activist visions for Japan's geopolitical future. Both of them are highly critical of current domestic politics and advocate thoroughgoing change. Both feel that Japan's economic power and technological capacity are valuable national resources that should be brought clearly to bear on the world stage.

The Realists are more pragmatic in their thinking, arguing that Japan, as it expands its international political presence, cannot avoid taking account of prevailing global circumstances, particularly the deep existing Japanese interdependence with the United States. Representative of the Realist viewpoint is the influential thinking of Ozawa Ichirō, cofounder of the Shinseitō (Renaissance) Party, architect of the anti-LDP coup d'état of 1993, and secretary-general of the New Frontier Party (Shin Shintō) that ultimately rose in its wake, during late 1994. Profoundly activist by both temperament and philosophy, Ozawa was scathingly critical of Japan's inaction during the Gulf War, when he served as LDP secretary-general. He fervently pressed legislation enabling the dispatch of Japanese forces overseas under the UN banner, failing during the Gulf War, but succeeding months later in securing the postwar dispatch of minesweepers to the Persian Gulf, as well as the subsequent participation of Japanese peacekeeping forces in Cambodia during 1991–92.

Ozawa's thinking is laid out in his *Blueprint for Building a New Japan (Nihon Kaizō Keikaku)*, which sold 700,000 copies within a year in its Japanese-language original.[34] He does not advocate straightforward rearmament or constitutional revision to permit it. But he argues that Japan must become a "normal country" (*futsu no kuni*) that "willingly shoulders those burdens regarded as natural in the international community. It does not refuse such burdens on account of domestic political difficulties. Nor does it take action unwillingly as a result of 'international pressure.' "

Ozawa is quick to note that the only overseas use of force that he foresees for Japan is under the flag of the United Nations. Yet he advocates a substantial expansion of peacekeeping involvement beyond what Japan did in sending minesweepers to the Middle East (1991), an engineering battalion to Cambodia (1991–92), an observer detachment to Mozambique (1993), and SDF medical assistance to Rwandan refugees (1994).

Classical analogies permeate Ozawa's rationale for direct Japanese military involvement abroad, as he rebuts the view that Japan should be a "trading nation" without military capacity. Venice, he notes, "did not survive 1,000 years simply because of superior business practices. It was a fully functioning republic: Venetians engaged in political and security efforts."

Carthage, by contrast, is for Ozawa the model of what to avoid. Enduring for six hundred years before being destroyed by Rome, it "offers a rich illustration on how to perish. Unlike Venice, it paid mercenaries to defend it. Its belief that wealth alone could sustain a nation ultimately caused its demise."

The reference to Japan's $13 billion contribution in the Gulf War, in place of any sort of military involvement, is veiled but unmistakable. Next time, if Ozawa has his way, Japan will be more proactive. It will not be inhibited in its geostrategic response simply by preexisting practice. It will not simply pay others, including the United States, to shoulder its defense burdens. It will act.

One central focus of Japanese activism, Ozawa suggests, should be East Asia. Such activism should be prefaced, he emphasizes, by a forthright recognition of Japanese war responsibility. "We have to admit that our government has not made much effort to settle the past. Nor was public feeling sufficiently harsh to prevent the reemergence of politicians associated with Japan's earlier aggression."

Beyond remorse, however, Ozawa prescribes intensified diplomatic initiative. "We must seek to define Japan's role and responsibility for building a stable regional order through discussions with the various nations of the region. It is precisely this sort of affirmative effort that will earn Japan the trust of the region."

Parallel in most respects to the political Realism of Ozawa, who stresses the realization of Japan's power potential through a more activist geopolitical partnership with the United States, is the military Realist school. Its point of departure is Japan's military rather than international political requirements.[35] Japan, the military Realists suggest, should develop strategies to meet the most likely military threats, not restricting strategic thinking to options that fall within the domestic political constraints against a major defense buildup. They, theorists differ from the Gaullists in not calling for security policies independent of the United States.[36]

Military Realists do, however, think more independently of existing constitutional constraints than do the more circumspect political Realists. Indeed, some of them, such as former Japanese ambassador to Saudi Arabia Okazaki Hisahiko, elaborate at length on how constitutional interpretations have distorted Japanese security policy and deprived it of any systematic relationship to strategy.[37] Among the political intrusions they cite are weapons limitations (no ICBMs, B-52–type long-range bombers, or even in-flight refueling capacity), nuclear inhibitions, the arms-export embargo, and the long-standing 1-percent-of-GNP limit on defense spending.[38]

From the military Realists' perspective—directly opposite to Progressive thinking—enhanced Japanese military capacity is the best possible way to avoid war. Powerful prospective adversaries, they argue, will naturally prey on a lightly armed nation. Most military Realists see the Anglo-Saxons and the Russians as the dominant prospective powers of East Asia and the Russians as natural adversaries, since Japan geographically blocks Moscow's four straits of access to the open Pacific. Such reasoning has caused the military Realists to continue exhibiting a strongly anti-Russian bias even with the collapse of the Soviet Union and the waning of the Cold War. This orientation is based on geostrategic thinking rather than just resentment of Russia's continued occupation of the Northern Territories.

The fourth school of thinking about Japan's geopolitical future is Gaullist. Like Charles de Gaulle himself in the 1960s, the Japanese adherents of this school, such as longtime right-wing LDP leader Ishihara Shintarō, profoundly doubt America's commitment to Japan.[39] Emphasizing cultural and ethnic differences, they are also skeptical about any presumed long-term community of interests between the two nations. The implication, as Ishihara argued in his 1989 one-million-copy best-seller *The Japan That Can Say No*, is that Japan should pursue a more assertive, nonaligned diplomacy among the major powers, consciously employing its economic and technological power for geostrategic leverage.

Since Japan produces the computer chips needed by the Pentagon, Ishihara observed, the United States would be totally helpless if Japan refused to supply them and sold them to the Russians instead. Japan should thus feel confident in saying no to America in the many instances—U.S.–Japan aviation negotiations, for example—when its demands are unfair.

The logical security implication of the Gaullist nonaligned stance, of course, is the need for a Japanese military buildup. If Japan cannot or should not rely on others, it must stand up for itself. Virtually all the Gaullists favor eliminating political constraints on defense policy, including the ban on arms exports, the three nonnuclear principles, and arbitrary GNP-based constraints on defense spending. Some also feel Japan should have nuclear weapons. The military doctrine running through this Gaullist thinking is emphatically one of independent deterrence.

The Emerging Outlines of National Consensus

Japan clearly is deeply divided regarding its future security stance. Its internal divisions have been graphically clear in the bitter, impassioned debates since the Gulf crisis began over dispatch of peacekeeping forces overseas. These have also provoked renewed calls for intensely controversial revision of the constitution, an issue that has surfaced again seriously in public discussion for the first time since the late 1950s. Yet despite the bitter divisions, some outlines of possible future national consensus are beginning to emerge.

PRIORITY TO ASIA

The most important of these is a rising precedence to Asia in Japanese strategic thinking. For the Traders, the reality since 1991 of trade volumes between Japan and Asia that are larger than those with the United States naturally supports this priority, which Progressives, with their bias toward mainland China, have long shared. For the Gaullists, a new Asia card is a welcome step toward more equidistant diplomacy—a way of forcing the U.S.–Japan relationship farther into the background.

The early focus of Japan's revived Asian interests, beginning around 1987, was the Asian newly industrializing economies (NIEs), particularly South Korea and Taiwan.[40] Business magazines and the mass media, especially television, gave them unprecedented coverage, most of it markedly more favorable than in the past. The elaborate Korean preparations for the Seoul Olympics of 1988 and the positive implications of yen revaluation for increased Korean and Taiwanese trade competitiveness were two major themes.

In 1988 this focus on the Asian NIEs broadened to include ASEAN. Just before Prime Minister Takeshita Noboru left for the Toronto G-7 summit that year, the national NHK TV broadcasting network ran an elaborate three-part series on the Asian Pacific (the NIEs plus ASEAN.) Throughout the year a range of publications stressed Southeast Asia's rapidly improving manufacturing plants—many of them, of course, Japanese-affiliated.

The deepening, largely favorable media coverage of Asia in the late 1980s coincided strikingly with new policy initiatives toward Asia by the bureaucracy and the political world, which in turn followed a surge in private-sector investment provoked by the rising yen. In 1987, MITI announced its new Asian Industries Development (AID) plan, designed to relocate Japanese industry into lower-cost Asian countries. In 1988 the Economic Planning Agency recommended a comprehensive integration of the economies of Asia, with the Japanese bureaucracy to play a catalytic role. At the Toronto G-7 summit of 1988, Prime Minister Takeshita Noboru explicitly chose to raise the concerns of the Asian NIEs and ASEAN, acting as an overall Asian representative to the highest council of the industrialized nations.

Feeding the rising emphasis on Asia in Japanese thinking, while simultaneously increasing pressures toward Gaullism in Japanese foreign policy, are the extensive informal activities in Japan of Malaysian prime minister Mohammed Mahathir. While Mahathir has not come to Japan as an official state guest since 1983, he has made innumerable personal visits, including many to Japanese *onsen,* or hot-springs resorts. Mahathir also comes to shop—especially at Tokyu Hands, a well-known retailer in Tokyo's Shibuya Ward catering to do-it-yourself hobbyists.[41] In his frequent trips to Tokyo, Mahathir has developed an extensive network of local mass media and political contacts, through which his caustically critical views of the U.S.–Japan alliance have come to have increasing impact on Japanese mass opinion.

In 1994 alone, Mahathir coauthored two provocative books sharply critical of Japan's traditional relationship with the United States, each with a best-selling Japanese author. In *The Asia That Can Say No*, coauthored with the Gaullist Ishihara Shintarō,[42] Mahathir stressed that Japan belongs naturally in the Asian geopolitical community, while the United States does not. He also questioned U.S. military security guarantees in Asia, arguing that the best security for Japan is to maintain friendly relations with Asia. To this end, he strongly recommended joining the regionalist East Asian Economic Caucus (EAEC), a grouping he has proposed that pointedly excludes the United States.

Mahathir also, like Ishihara, urges Japan to play a much more proactive international role, rather than only reacting passively to American suggestions. He argues that Japan should take its "natural place" as a world leader by joining the UN Security Council as a permanent member and shedding inhibitions about sending its troops overseas. In his second 1994 book, *The Asians and the Japanese*, with former McKinsey Tokyo Bureau chief Ohmae Kenichi, Mahathir also stressed that Japan should stop apologizing for World War II, devoting its energies instead to actively aiding Asian development.[43]

Despite Asia's clearly rising priority in Japanese thinking—stimulated by Mahathir, Lee Kwan Yew, and others—many Japanese clearly retain some of their nation's traditional ambivalence. This, in turn, is stimulated and intensified by their neighbors' continuing bitterness against them. Only 41 percent of Japanese respondents, in a July 1995 *Asahi Shimbun–Dong A Ilbo* joint poll, for example, thought Japanese–Korean relations were going well, and only 13 percent of Koreans thought so. Only 11 percent of Japanese said that they liked Korea, and only 6 percent of Koreans said they liked Japan, with negative sentiments much more pronounced, especially on the Korean side.[44] There clearly remain limits on Japan's "return to Asia," especially in the northeastern part of the region.

QUIET AMBIVALENCE ABOUT CHINA

While the four schools of Japanese strategic thinking are virtually unanimous about the importance of deepening Southeast Asian and, to a slightly lesser extent, Japan-Korea ties, the steady rise of China stirs a more complex reaction. China was long the traditional focus of Japan's ties with Asia and was its historic cultural mentor. Virtually every Japanese can recite lines and lines of Chinese poetry by heart.

Most Japanese intellectuals find it difficult to criticize China, for many of the same reasons that Germans are reluctant to criticize Israel. The heritage of twenty million Chinese deaths in Japan's long history of aggression against China continues to loom large. Yet many Japanese, particularly those close to Taiwan, quietly resent what they consider a series of unwarranted, imperious Chinese actions. Among their complaints are the 1974 show of force around the disputed Senkaku Islands; China's continuing efforts, as they see it, to keep Japan out of the UN Security Council; China's persistent nuclear testing; and China's abrupt 1992 reassertion of its territorial claims to large stretches of the East China Sea.

Continuing Japanese repugnance at the 1989 Tiananmen massacre, together with criticism of continuing human rights violations in China, is less pronounced than in the United States, but nevertheless significant, especially among the Progressives. More important in the pragmatic Japanese international affairs calculus are quiet misgivings about rising Chinese economic and military strength. The Realists and the Gaullists, of course, feel this most keenly, but Chinese mercantilism and systematic favoritism for overseas Chinese interests over those of Japanese firms has begun provoking many Traders as well.

The rapid economic rise of China seems to be having a much stronger impact on Japanese perceptions and expectations for the future—both of leaders and the general public—than is true in the United States. A mid-1995 Nikkei–Dow Jones poll found, for example, that 16 percent of Japanese already regarded China as the strongest economic power in the world, compared to 5 percent of Americans. Looking to the future, a full 66 percent of Japanese saw China as the strongest economic power by 2015, compared to only 17 percent of Americans. Twice as many Japanese as Americans also saw China as a threat to world peace in the future, although this was distinctly a minority view.[45]

Japanese perceptions of rising Chinese economic strength are combined with lingering doubts about that giant country's prospective political stability and apprehension regarding the broader regional implications thereof. Tokyo University international relations specialist Tanaka Akihiko calls the situation "peaceful but unstable."[46] He considers the fact that China is not currently engaged in major military conflict extremely significant, but also notes that most of the destabilizing factors in East Asia are linked to China.

The skeptical, somewhat apprehensive dimension of Japanese views toward China is amplified in the work of well-known China scholar Nakajima Mineo of the Tokyo University of Foreign Studies.[47] Nakajima stresses the danger that uneven development within China will cause internal disparities to become too great, with destabilizing consequences. He is also concerned about how the immense economic and political power of the "three Chinas" (China, Taiwan, and Honk Kong) can be integrated into the Asia of the future in stable fashion. If unified, this dynamic trio, projected by the World Bank to have a purchasing-power parity-adjusted GNP larger than the United States by 2002, will be a power threatening to Japan. But division and conflict among them will create different instabilities of its own, he maintains.

Many Japanese, in sum, are left with a vague sense of foreboding at emerging trends in China. They see the rise of a new economic giant—in much more stark and magnified terms than do even Americans. Yet at the same time many Japanese are impressed by the prospective turbulence of that huge land, only a few hundred miles from their own vulnerable shores. Despite their traditional lack of security consciousness, it would not be unreasonable to expect that they might search seriously for new means of either balancing or co-opting China in coming years.

SKEPTICISM ABOUT "BUSINESS AS USUAL" WTH AMERICA

Related to the rising emphasis in Japan on Asia, and a broad intellectual trend stressed by many of the same people, is a skepticism about the relative importance of U.S.–Japan bilateral ties. Once again, Traders, Progressives, and Gaullists can all agree on this, as Japan's economic relationships gravitate steadily toward Asia, as grass-roots frictions relating to the bases intensify, and as the United States fails to offer attractive reasons for reviving a Washington–Tokyo entente. Unremitting criticism from Washington of U.S.–Japan economic relationships that many Japanese have come to see as less and less intrinsically valuable only intensifies a Japanese impulse to diversify political and economic links still further. This was clear in Japanese response to the Clinton economic diplomacy of 1993–94 and 1994–95, which encouraged Japan to deepen its links with China, Indonesia, Malaysia, Vietnam, and other major East Asian states. Populist resentment of the bases also provoked suggestions of top LDP leaders such as Katō Kōichi in late 1995 that Japan should "rethink" peripheral elements of the U. S. military presence in Japan, such as the large amount of airspace and telecommunications frequencies allotted to U.S. forces.

One particular line of emerging emphasis in Japanese thinking, as it marginally diversifies away from its long-standing U.S. focus, is greater stress on the United Nations. Expanded UN peacekeeping roles, for example, were a central concern of the landmark 1994 advisory committee report to the prime minister on post–Cold War security, chaired by businessman Higuchi Hirotarō. This rising UN emphasis in Japan, and veiled ambivalence about narrowly U.S.-centric security ties, runs in tension with currents of unilateralism rising strongly in the U.S. Congress and in broader American public opinion.

There seems to be, in sum, an emerging consensus among several schools of thought in Japan on the virtues of gradual disengagement from the American embrace. Apart from this, a shift in the relative influence of the four schools of thought outlined above is also worth noting. Most strikingly, the influence of the Progressive school has fallen substantially, with a sharp about-face by the Japan Social Democratic Party regarding the legitimacy of the Self-Defense Forces. Long the SDF's most caustic foe, the Socialists during the summer of 1994 suddenly shifted their stance and accepted its legitimacy, as part of a political deal by which their leader Murayama Tomiichi became prime minister. This sudden development shifted the entire fault line of Japanese security discourse significantly to the right, reviving prospects for constitutional revision and an even more forthright discussion of how Japan should project its rising political influence abroad.

Absent a potent external shock to set a new course for national policy, Japan seems unlikely in the balance of this century to radically realign itself in international affairs. The ongoing internal struggle for strategy is too bitterly divisive to permit such a decisive conclusion. Protracted recession during the first half of the 1990s, combined with a series of tragic and sometimes bizarre shocks like the 1995 Kobe earthquake and several subway poison-gas attacks, temporarily deepened the pessimistic, quietist mood of many Japanese.

Yet the same divisions that inhibit short-term activism are also likely to frustrate efforts to actively constrain more troubling long-term forces that even now quietly corrode the status-quo framework of U.S.–Japan relations. Japan's dependency on and orientation toward Asia—where it now has well over $60 billion in direct investment—will almost inevitably rise. And its sensitivity to American interests will decline. Japan's military will become both more technologically proficient and more independent of America. How might this new giant, torn from its transpacific moorings, respond to the turbulent Asia so close to its placid shores?

THUNDER OUT OF CHINA

After hours of switchback, nighttime driving across the barren, windswept hills of southern Yunnan, the People's Liberation Army convoy slowed for the Salween River bridge. Screeching into low gear, its heavy trucks, laden with munitions and spare parts for rocket launchers, mortars, and armored personnel carriers, rumbled on down the Burma Road toward Wanding Ruili, and then Lashio beyond. Reaching the frontier before daybreak, they were whisked through the Burmese checkpoints without fanfare on their inexorable course to the south. In the moonless darkness they passed in low gear through border boom towns with their karaokes, massage parlors, and discos now quiet for a few hours before dawn— the symbol of Burma-Yunnan border trade that has spiraled thirtyfold to nearly \$1 billion in the past decade.

Fifty years ago, American Seabees built this Burma Road, to funnel Lend-Lease supplies northward to Chiang Kai-shek's halfhearted struggle against Japan. Today, in the dawning age of a powerful, prosperous—and nationalistic—new China, it is the dominant path for Chinese military and economic aid to a newly strategic ally.

Myanmar[1] has few energy resources. Exploitation efforts in the Gulf of Martaban have come up largely dry so far. But as China becomes a major oil importer, Myanmar, lying between China's Yunnan province and the Indian Ocean seaways to the Persian Gulf, assumes new and rising geostrategic importance.

China has not expanded into Myanmar in any violent way. It has provided welcomed consumer goods and extensive aid for the dominant local military. Yet its presence is nevertheless intimidating to many in a nation with a tenth the number of men under arms, a twenty-seventh of China's population, and a thirty-seventh of its GNP.

China has used its leverage strategically. In late 1992, Western spy satellites, for example, detected a new 150-foot antenna used for signals intelligence at a naval base on Coco Island, a Burmese possession on Indian Ocean sea routes less than a thousand kilometers northwest of the Strait of Malacca. Western analysts believe Myanmar is allowing Chinese technicians to operate this as a listening post. More recently, China has been pressing Myanmar to allow access to Victoria Point, a long, rugged Burmese island within three hundred kilometers of the Strait of Malacca, the vital seaway through which much of Northeast Asia's trade must pass.[2] China is also upgrading the Burmese navy, together with the roads and railroads that lead from its Yunnan province southward toward the Indian Ocean. The Chinese press regularly refers to Myanmar as part of "the Great Golden Peninsula," from Yunnan to Singapore in the south, India in the west, and Vietnam in the east. That new frontier clearly offers major, at least commercial, opportunities for the future.

Across the long sweep of its four-thousand-year history, China rarely has been aggressively expansionist. It has never had a Hitler or a Napoleon. And there is no particular reason to expect one now.

Yet a complex domestic equation, with factors including growth, energy shortage, domestic social transformation, and generational leadership change, is causing China to project outward in unsettling new ways. It is that equation that concerns us here. The forces at work are those that power the PLA's convoy rolling south along the Burma Road, relentlessly, toward the sea-lanes and the vital energy supplies that they convey in ever-growing quantities.

China's Growth and the Geopolitical Equation

Western analysts have often been naively optimistic about China's growth prospects, for two hundred years and more. Yet both technical analysis and recent experience suggest that this time the current impressive growth spurt may persist. China has averaged, under diverse political conditions, more than 9 percent real economic growth ever since its reforms began in 1978; its factor endowments and cost structure are such that 6 percent growth could continue for at least another generation. If additional fiscal, financial, and regulatory reforms are achieved, growth to the year 2000 might even match the 9 percent average of 1979–94.[3]

High rates of savings and investment underlie the buoyant short- and midterm prospects for China's economy. Total household bank savings, for example, were 46 percent of GNP in 1991.[4] Demographic trends, including a generally declining birthrate since the mid-1970s, could help sustain these levels for another generation. China's vast reservoir of manpower for labor-intensive industries, coupled with the huge size of its domestic market, should also continue to attract substantial foreign investment. Even with substantial attrition from the $110.9 billion in foreign investment approved during 1993, there is a huge amount of investment capital in the pipeline to supplement already high domestic savings across China. Actual investment in 1994, for example, normally far lower than more tentative approvals, reached $35 billion, more than that entering any other nation except the United States.[5]

Total factor productivity in China has already risen substantially over the past fifteen years, especially in the countryside. But it still remains low in comparative perspective, boosting growth prospects for the future. The level of literacy in China is unusually high for a developing nation: around 73 percent, compared to 54 percent in Iran and 48 percent in India, for example.[6] Chinese vocational training is also broadening and improving rapidly. Infrastructural bottlenecks in electric power, communications, and transportation are steadily being overcome. China already, for example, has on the drawing boards plans for a nationwide network of bullet-train lines, similar to that of Japan, linking major urban centers across the country. France and Japan are both vying to build it, pitting the latest Shinkansen and GTV technology against one another in an epic contest.

No doubt there are limits to the volume of Chinese exports that the world can politically and physically accept, even if competitively priced. But those constraints have by no means yet been reached, with Chinese trade remaining far smaller than the overall size of the economy. Chinese merchandise trade in 1993 was about 2.5 percent of global totals, just surpassing China's share of the late 1920s.[7] Yet China's domestic economy accounted for 6 percent of world output, or more than double its still-meager share of world trade. China's domestic market is so vast that even if foreign export restrictions intensify, domestically driven growth could well continue for decades thereafter, as economic development gradually broadens away from the coast into the vast interior.

The long-term geopolitical implications of Chinese growth, of course, remain to be seen. Chinese military spending increases have more than equaled China's explosive economic growth of late, even as other major powers have cut back.

And the military has gained major additional support from as many as twenty thousand military-run commercial enterprises around the country.[8] Yet the initial technical level of Chinese forces, before modernization began in 1978, was very low, especially of the navy and air force.

Is China just playing catch-up? How serious are its aspirations to superpower standing, and how threatening are they to other nations? The answers are unclear. Yet China's population, area, natural resources, military capacity, and now economic scale give it potential for independent great power standing that no other nation in Asia, including the Japanese economic colossus, can match. The growth now occurring in China could well continue to be translated into military potential, and to broader Chinese influence in world affairs, although the form and pace of such changes are highly uncertain.

Why China Will Not Be the Soviet Union

The Soviet Union too, of course, was once a huge, underdeveloped continental power—almost three times larger, in geographic terms, than even China. Yet it collapsed in chaos, both economic and political, with sweeping global geostrategic implications. Despite the myriad critical uncertainties of China as it approaches the new century, it seems quite clear that China will not follow Russia's chaotic transition. Deng Xiaoping has not been China's Gorbachev, and none of his successors is likely to be either.

For all the superficially common socialist heritage, underlying differences in political system and social structure are helping China avoid the Soviet Union's fate. The USSR was an ethnically plural federation state—only about half Russian—with severe internal ethnic divisions that led it ultimately toward dissolution. China, by contrast, is around 93 percent Han Chinese, with much greater natural ethnic cohesion. Because of the nature of its mass-based agrarian revolution and Mao's emphasis on egalitarianism, its government, it could be argued, has been more responsive to the grassroots. In Moscow, by contrast, Lenin took power in 1917 via a Leninist coup brutally imposed from above. Like the French and most Latin American dictatorships, Russian government has always had problems responding to pressures from below. These difficulties were compounded by the abstract intellectualism of Gorbachev, which contrasted sharply to the pragmatism of Deng Xiaoping.[9]

Flowing from differences in revolutionary tradition and interest-group structure, the Chinese regime established an early and deep base of support in rural areas and many urban localities. That populist backing has greatly aided stable, rapid economic growth, just as analogous if nonrevolutionary coalitions did in

Japan, South Korea, and Taiwan.[10] Favorable agricultural prices for farmers were a constant priority of the Maoist regime, even as it radically collectivized the countryside through the commune system.[11]

Deng Xiaoping's first initiative was likewise farm reforms, this time in an unambiguously market-oriented direction. His reforms quickly doubled the income of China's 800 million farmers and thus established a firm political basis for his broader program. Deng expanded his coalition by backing the rise of small-scale entrepreneurs. He also brightened grassroots employment prospects by support for labor-intensive light industry.

Gorbachev, by contrast, neglected farmers and entrepreneurs, in Lenin's and Stalin's tradition, thus narrowing his political base. And he made things worse by banning vodka. He also darkened prospects for employment gains, through a focus on heavy industry that generated high-paying, high-value-added jobs, but only small numbers of them.[12]

Another fundamental cause of Gorbachev's failure was his strategy of reform: attempting perestroika in every sector at once, as Poland also did in its "Big Bang" reforms of the early 1990s. As in the case of the shah of Iran's reforms in the late 1970s, this approach antagonized so many groups simultaneously that the Soviet political system was overwhelmed by opposition, and chaos ensued. China is not trying this sort of high-wire act.

Unlike Gorbachev, Deng Xiaoping and his successors have pursued a very cautious, gradual, sequential approach. They began, as indicated above, with politically riskless rural reforms that both doubled peasant income and put more food on urban tables. They then systematically broadened the scope of reforms, focusing on areas where limited government investments would produce rapid growth. Unlike the Soviets, they enthusiastically welcomed foreign investment, which created jobs, provided technology, and steadily expanded the tax base.

Unlike the Russians, the Chinese have been rather slow to privatize, waiting for broader institutional and macroeconomic conditions to emerge that could optimize the success and efficiency of such firms once they entered the private sector. Thus China hopes to avoid the dual dangers of monopoly profiteering and bankruptcy that have often beset Russian firms privatized while broader price distortions in the economy are still pervasive. China has placed more stress than Russia on the development of financial markets, especially a bond market, that can absorb inflationary pressures, amass capital for investment, and ease the privatization process.

There is one last Sino-Soviet contrast that must be noted. China has, throughout its reform process, been much more integrated into the global political economy than was Gorbachev's Russia, its post-Tiananmen isolation notwithstanding. In 1990, trade constituted 25.4 percent of GNP in China, compared to 16 percent in the United States and only 5 percent in Russia.[13] International capital flows to China are also much greater than to Russia. Beijing had a Hard Rock Café long before Moscow did.

China, as a whole, was from the beginning a more cohesive and unified nation than the Soviet Union. It has also pursued, the tragic 1989 Tiananmen massacre excepted, more politically astute developmental strategies than Russia and received more international economic support. To be sure, China confronts troubling political and social challenges including wholesale repression of human rights, that may yet generate instabilities of their own. Yet its reforms by and large stand on a stable foundation, in contrast to those of Russia, making Russian-style economic collapse in China highly unlikely.

Social Tensions Amid Rising Affluence

The likelihood of rapid, prolonged growth, however, does nevertheless give rise to some troubling domestic social and political prospects. These could well help give Chinese foreign policy a more assertive cast, as leaders desperately strive to unify their people. At the heart of the problem is the massive rural population of China, now over 800 million strong. Because of rapidly rising agricultural productivity in recent years, many Chinese peasants have little to do in their home villages. In Mao's China they were effectively controlled by their local communes, until those were gradually dismantled by Deng Xiaoping in the early 1980s.

Rapid economic growth, centering in the coastal cities, has given rise to sharp interregional income differentials. Inland Hunan and Guizhou provinces in south-central China, for example, have per capita incomes one-half and one-third respectively of levels in booming Guangdong province adjoining Hong Kong.[14] During Maoist days (1949–76), these regional differences were minimal. And the peasants of inland areas are even poorer, relative to their coastal urban cousins, than province comparisons alone might suggest.

Rising income gaps, a commonplace of economic development worldwide and a frequent source of political instability, could be especially dangerous in China. The close identification of the current government with the economic development process, coupled with rising apparent corruption of government officials themselves, would make this so. Such inequality can also be more easily laid than in most countries at the doorstep of government, since China is, after all, at least nominally a socialist state. China's revolutionary heritage—intensely egalitarian—also provides a visible and broadly familiar critique of the deepening social pathologies, bred by the high-growth process, that Chinese can now see unfolding before their eyes.

As communal controls collapse in the countryside, the natural consequence of growth and rising income differentials is a flood of migrants to the cities. Ten years ago this would have been stanched at the urban end by *danwei* (state work units), or by street committees monitoring all those formally lacking a local residence. Yet the power of these officious bodies is also being eroded. Economic growth and poor state salaries lead workers to multiple jobs, and declining reliance on any one particular workplace.

China today has an estimated 100 million illegal migrants from the countryside who sleep in front of railway and bus stations, living on everything from panhandling to odd jobs to selling vegetables in tiny stores. With economic growth and rising rural expectations, their numbers are rising sharply. In Guangzhou alone more than a million migrants—a sixth of the entire population—lack residence permits or stable work. Wherever one walks in the center of town one meets them—leaning on their meager possessions, selling pineapple at roadside stands, or hawking anything from cigarettes to winter coats to J. Crew sweaters. The situation is similar in Shenzhen, Beijing, and Shanghai.[15] As long as China's current economic boom continues, the great cities can absorb this "blind stream" flowing toward them from the countryside. But if growth slows, unemployment and urban social tensions will almost surely rise.

The erosion of long-standing Communist institutions that provokes large-scale migration is an urban as well as a rural phenomenon. Although now producing less than half of gross national output, China's mammoth state enterprises remain grossly overstaffed, and heavily subsidized. Well over one-third of them are losing money; their acknowledged losses exceed the national budget deficit and amount to 4–5 percent of GNP, according to the World Bank.[16] They are attracting a growing share of government investment capital—70 percent in 1993, as opposed to 61 percent in 1989.[17] But this change is, sadly, due more to their inefficiency and political ties than to any innate strategic importance or economic promise.

Bloated state-owned factories employing 100,000 people and supporting whole cities are common. The Wuhan Iron and Steel Company, for example, a thousand kilometers inland from Shanghai on the polluted banks of the Yangtze River, has 120,000 employees and is virtually the sole source of livelihood for 300,000 people. Driving for miles through the acrid surroundings of the plant, one passes forty-seven schools, a two-thousand-bed hospital, and 2.4 million square meters of drab Stalinist apartment buildings—all financed and supported by this struggling local steel mill.

The Wuhan plant and its surroundings epitomize the "iron rice bowl" of unqualified cradle-to-grave security that was urban socialist China under Mao Zedong. But the grinding pressures of the market are now stressing that iron rice bowl to the cracking point. The central government pours a full seventh of its total revenues into subsidies for such firms[18] and supplies much larger sums in loans from state banks, which will never be repaid. These increasingly threaten the viability of the national banking system. About 41 percent of all companies in the state sector record losses, which rose around 28 percent in 1994.[19] An estimated 68 million urban Chinese will be looking for work, largely because of retrenchment in such government firms, over the next decade alone.[20]

Despite a one-child reproduction policy, China's population—already one-fifth of the human race—is still growing by fourteen to sixteen million a year. By the year 2000 it is expected to exceed 1.3 billion; and by 2050, 1.6 billion. With such population growth, unemployment in the countryside—where the policy is more difficult to enforce—will probably soar; a government commission recently predicted 300 million "surplus" workers there by the year 2000. Rationalization of state enterprises in the cities will simultaneously intensify, even as the population shift to the coastal areas accelerates. The specter of pervasive national unemployment and volatile migrations across China is clearly looming, especially should economic growth fall below expectations.

The migration problem is provoking an upsurge of crime in its wake. Crime rates have already tripled in the fifteen years since Deng Xiaoping's reforms began, despite tough Chinese deterrent steps that include nearly two-thirds of the executions carried out in the entire world.[21] This wave of violence threatens to intensify as migration and unemployment both steadily expand, in tandem.

China's deepening migration problem, rooted in social change, economic growth, and the inexorable pressures of people on land, has implications not only for domestic stability but also for international relations. It raises the specter of huge refugee flows that could compound unemployment and welfare burdens throughout the Pacific. Most immediately affected is Russia, whose sparsely populated Far East region, with only eight million inhabitants, is dwarfed in population by neighboring Manchuria, with over sixty million. Tough visa restrictions clamped on Chinese visitors in early 1994 can only temporarily stem the tide which has already, by some reports, brought close to one million Chinese squatters to the Russian Far East.[22] Over 100,000 Chinese are also believed to be illegally entering the United States every year, either by land, air, or sea,[23] with many thousands more overstaying their visas in Japan every year. Hong Kong and Taiwan, of course, also seem likely to confront waves of illegal immigrants as their ties to China deepen, especially in the ambiguous conditions likely to follow Hong Kong's return to the Chinese mainland in mid-1997.

The Slipping Hold of Beijing

Together with rapid growth and the threats to social cohesion that it provokes, another predictable trend in China over the coming decade or so is rising political decentralization. China, to be sure, has been a more or less unified nation-state ever since the reign of Shih Huang-ti, who built the Great Wall two centuries before Christ. Its people are more than 90 percent Han Chinese, as noted. But China's regions have much stronger independent cohesion and legitimacy of their own today than is often appreciated.

Indeed, effective central control of the provinces may already be lower than at any time since the introduction of central planning in the mid-1950s.[24] Beijing tried to co-opt rising provincial power, rooted in the explosive economic growth of the early 1990s in China's coastal area, by adding six provincial leaders, from Shanghai, Shandong, Guangdong, and Tianjin, as well as Tibet and Beijing, to the Politburo at the Fourteenth Party Congress in 1992. But under virtually any scenario for national leadership succession, those provinces seem likely to grow even more powerful in coming years.

Resurgent regionalism has a strong historical and cultural basis in China. Provincial boundaries have been largely set since the eighth century, and well-defined regional stereotypes based on those boundaries have existed since at least 1200.[25] It was not until the twelfth century that southern China ("the land of the southern Sung," to the south of the Yangtze) was effectively conquered by Han migrants from the north. Manchuria was only really assimilated in the nineteenth and early twentieth centuries, while control over Sinkiang and Tibet is even more recent. "Core China"—that assimilated more than two hundred years ago—is only 45 percent of the current China's total geographical area. And newly acquired Sinkiang and Tibet—sprawling, geostrategically important areas on China's sensitive western frontiers—both have restive majorities that are not Han Chinese, 55 percent in Sinkiang and a full 95 percent in Tibet.

Even within "core China," cultural and even physical differences are legion. Chinese speak six major regionally based dialects apart from standard Mandarin, which are as mutually incomprehensible as Gaelic and English.[26] Northerners remain on average five centimeters taller than Southerners. Chinese from the south and the coastal provinces are strikingly more commercial by tradition than their more austere and bureaucratic brethren in the north.

The simultaneous growth, economic liberalization, and deepening interdependence with the world now occurring in China is steadily eroding the power of the center. In part, this change is a conscious devolution, initiated by Deng Xiaoping in 1991 to outflank conservative opponents of economic reform in the Beijing *nomenklatura*. But devolution has fed on itself, spurred by the natural desire of local authorities in the affluent and increasingly powerful coastal provinces to appropriate more and more of the fruits of growth to themselves alone. By 1993, Beijing's revenue had fallen to only 40 percent of the national total, with the remainder in the hands of the provinces. This central government share was only 19 percent of GNP, down from 34 percent in 1978, as the reforms began.[27]

Beijing's declining ability to control tax revenues is disturbing in two respects. First, it tends to be inflationary, as it encourages higher levels of deficit financing. Second, it reduces the central government's ability to redress income and growth inequalities that are becoming ever more pronounced as growth proceeds. In 1991, for example, per capita GNP in Shanghai was 3.5 times the national average, twice the average in Tianjin, and 1.5 times the average in Guangdong. Conversely, over 60 percent of China's poor lived in the southwest. With limited resources, co-opted largely by defense spending on the one hand and subsidies to state enterprise on the other, the central government has grown ever less able to cope with looming inequities. These are dangerously eroding the legitimacy of the central government itself across much of China, which after all continues, at least in form, to call itself socialist.

The rising power of the regions is also reducing the ability of Beijing to enforce a coherent set of trade and regulatory policies across the country. This problem is parallel to the difficulties that American authorities faced under the Articles of Confederation two hundred years ago, before the federal Constitution was adopted. Both Hunan and Jiangxi provinces, for example, have blocked grain shipments to wealthy, booming Guangdong, hoping to prevent a rise in local food prices. In 1993, conversely, Guangdong tried to restrict the number of workers flowing in from outside the province, although it was unsuccessful in doing so. Jilin, Honan, Liaoning, and Hubeh all have regulations allowing only the sale of native beers, wine, laundry, detergent, bicycles, and televisions.[28] Beijing seemingly lacks the power to curb this sort of "regional protectionism" against interprovincial commerce.

South Chinese Regionalism

A special variant of regionalism that will have profound implications for both China and the world over the coming decade is rapidly evolving along the southeast coast of China. This is the emergence of a "South China Economic Zone" linking Guangdong and Fujian provinces of China with Hong Kong and, to a lesser but still significant degree, Taiwan as well. Driven by Hong Kong's impending reunification with China in mid-1997 and Macao's reversion two years thereafter, cross-investment and trade are exploding along the littoral of the South China Sea.

The level of economic interdependence is already impressive. Approximately 70 percent of all the commercial investment in China today (and virtually all of that in the Southeast) either originates in or flows through Hong Kong.[29] Hong Kong's own investment in China stood at around $50 billion in the mid-1990s, while the PRC probably had $20 billion invested in Hong Kong. Around two-thirds of Hong Kong manufacturing is located in the Pearl River delta, where Hong Kong companies employ over three million workers—more than they do in Hong Kong itself. Nearly one-third of all the Hong Kong dollars in circulation float about in Guangdong province. Hong Kong's trade with China, mainly with Guangdong, totaled nearly $100 billion in 1993, up from $5.7 billion in 1980.[30]

The economic basis is steadily being laid for even deeper integration between Hong Kong and Guangdong, anticipating 1997. Entrepreneur Gordon Wu's $12 billion superhighway between Shenzhen and Guangzhou was completed in July 1994, shortening travel time from six to two hours, with cross-border links to Hong Kong and ultimately Macao impending.[31] Telecommunications and aviation links are also deepening rapidly.

Taiwan is also being drawn deeply into this South China economic network, despite still profound political differences with the mainland. Its trade with Hong Kong, much of which is ultimately destined for China, rose from $2 billion in 1980 to $17 billion by 1992. Meanwhile, Taiwan's trade with the mainland soared from around $300 million to $7.4 billion, rising further to $10 billion in 1993. Most of this rapidly growing trade was concentrated around the Fujian coastal areas with which many native Taiwanese have ethnic and linguistic ties, although during 1993–94 economic exchange began gradually broadening into the interior.

Investment flows among Hong Kong, South China, and Taiwan also began growing sharply in the early 1990s. By 1995, Taiwan had over $20 billion invested in China and 15,000 projects there, and more than $2 billion additional in Hong Kong. Its investment in Fujian was generating 70 percent of the industrial output in that entire province.[32] Hong Kong, as noted previously, had over $50 billion in China and $12 billion in Taiwan, while China had $20 billion in Hong Kong. Thus, cross-investment among the three parts of "Greater China" totaled over $100 billion, with the bulk of this concentrated in the South China Economic Zone.

This vibrant emerging economic community owes little to Beijing. To the contrary, Beijing's intervention is broadly considered—and not only by Taiwanese—as a bane rather than a blessing. There is also an ethnic dimension: Cantonese and Fukienese have closer historical and linguistic ties with their brethren across the Taiwan Strait or the Lo Wu Bridge separating Hong Kong and Guangdong than they have with the north of China.

Inhabitants of South China have been consciously seeking to strengthen these local ethnic ties. Southerners are refurbishing their historical narratives, using archaeological evidence of the Chu Kingdom, an ancient civilization flourishing in the south centuries before Shi Huang-ti's unification of all China, to reinterpret their cultural origins with a regionalist slant.[33] Similarly, native Taiwanese, whose language and folk rituals are very similar to those of people in southern Fujian, are restoring family temples and celebrating local gods in and around Xiamen, on the Chinese coast less than one hundred miles from their own shores.[34] Pop culture is also binding the people of the South China Economic Zone, as the Maoist fervor that once unified the PRC decisively fades. It is not Deng Xiaoping and his heirs but the memory of the late Teresa Teng, Taiwan's famous pop singer, that rules this part of China.

The Delicate Problem of Taiwan

Similar regionalist pressures to those corroding national unity in southern China are also operating in Taiwan as the generation of mainlanders that fled before the victorious People's Liberation Army in 1949 passes from the scene. They and their descendants, traditionally dominant on the political stage, now number less than 20 percent of Taiwan's 21 million people. Upon the death of Chiang Ching-kuo, Chiang Kai-shek's son, in 1988, a Taiwanese, Lee Teng-hui, succeeded him. In 1993 a mainland-born Taiwanese, Lien Chan, also succeeded an aged mainlander as prime minister.

Deng Xiaoping, as noted previously, identified five conditions under which he considered military force against Taiwan justified—among them a Taiwanese declaration of independence or "rejection of unification talks" for a long time. But 10–15 percent of the Taiwanese population, by most opinion polls, back a formal declaration of independence, despite Beijing's threats.[35] And the share approving of the Taiwanese independence movement rose steadily during the late

1980s and the early 1990s, from only 2 percent in November 1988 to nearly 24 percent in May 1993.[36]

Rising regional consciousness in Taiwan has been intimately linked to the steady growth of the opposition Democratic People's Party (DPP), established in 1986 in defiance of the martial law that had persisted on the island since May 1949. At first the DPP (known as the *dangwai* in Chinese) was generally regarded as a dangerous, eccentric fringe group, with its strong insistence on an independent Taiwan run by ethnic Taiwanese. Many feared that such claims would attract mainland intervention, or at the very least weaken international confidence in Taiwan's overall political-economic climate. But after martial law was lifted in July 1987, and as the media grew somewhat freer, the DPP began to score politically. This was especially true in the cities, where the DPP's emphasis on persisting social problems and the corruption of the ruling Kuomintang, not to mention their appeal to Taiwanese localism, proved most persuasive.

The DPP was encouraged by the breakup of the Soviet Union and the admission of the Baltic states and other former Soviet republics to the UN. In September 1991 the DPP launched a campaign for UN membership under the name of Taiwan, or the Republic of Taiwan. A month later, it also inserted a formal independence clause in the party platform. PRC president Yang Shankun warned, in response, that "those who play with fire will be burnt to ashes."

In the watershed 1992 parliamentary elections, the DPP won 31.9 percent of the vote, and together with other splinter groups denied the ruling Kuomintang (KMT) a two-thirds majority. With the KMT split into six major factions, the DPP thus gained considerable potential legislative power. A 1992 constitutional amendment providing for direct future election of the president promised to strengthen the forces of regionalism on Taiwan still further. So did the 1993 city and county governmental elections, where the DPP netted an unprecedented 41 percent and narrowed the KMT's margin to single digits for the first time. In December 1994 local elections, the DPP elected its secetary-general mayor of the capital city of Taipei.

The beginning of a semiofficial dialogue through front organizations between Taipei and Beijing in April 1993, with the first meeting held in Singapore, was an attempt to keep regionalism in check. Yet it has experienced only very limited subsequent progress. Low-level discussions have continued on such practical issues as hijacking, fishing disputes, and illegal immigration across the Taiwan Strait. Official representatives from the mainland have actually visited Taiwan. But clear-cut agreements have been long in coming. Low-level incidents such as piracy, airplane hijackings, accidental artillery fire, and the occasional murder of Taiwanese tourists on the mainland—not to mention China's increasingly elaborate and realistic military exercises in the neighborhood of Taiwan—continually complicate what efforts there are toward mutual accommodation across the Taiwan Strait.

Meanwhile, regionalist forces on Taiwan have continued to grow stronger. In March 1993 the Taiwan High Court acquitted Taiwan independence activist Chang Tsan-hung on charges of sedition. The founder and chairman of the World United Formosans for Independence was meanwhile allowed to move his controversial organization's headquarters to Taiwan.[37] Within Taiwan itself, views that Beijing considers dangerously subversive of long-run reunification prospects thus are vetted ever more freely.

Taiwan has also pursued a persistent, if often informal, government campaign for broader international recognition. By the early 1990s this had quietly yielded membership in the Olympic Games, APEC, and the Asian Development Bank. In May 1995 Taiwan engineered a major coup by maneuvering President Bill Clinton, through congressional pressure, into approving a visit by President Lee Teng-hui to speak at his Cornell University alumni reunion in Ithaca, New York.

This was the last straw for the Chinese. Determined to inhibit, and indeed destroy, an imagined U.S.–Taiwan conspiracy to move toward a two-China policy, they counterattacked vehemently on many fronts, arresting human rights activists, accelerating controversial Middle East missile exports, and holding ostentatious military exercises, including long-range missile tests, less than one hundred miles north of Taiwan. More than two hundred commercial flights in and out of Taipei were rerouted during the week-long missile tests.[38] Taipei's stock market dropped by more than 3 percent in a day and 30 percent over the first half of 1995, as Taiwan retaliated with its own military maneuvers and threats to reconsider its long-dormant nuclear program.[39] A best-seller on a fictional Chinese invasion of Taiwan, *August 1995*, sold 300,000 copies in a matter of weeks[40] and politicians on all sides moved to capitalize on the new spirit of resistance in Taiwan to Beijing's pressure tactics.

Uncertain Transition in Hong Kong

Meanwhile, in Hong Kong a vocal group critical of Beijing has also emerged since the 1989 Tiananmen massacres, which drew huge protests in Hong Kong. The movement has gathered support not only from students, journalists, and other traditional democratic activists, but also from large groups of middle-class citizens, especially professionals, who see their welfare threatened by the scheduled 1997 transition to Chinese rule. Catalyzed originally by the mass protests of over a million participants following the 1989 massacres in Beijing, the critics gained further momentum with the arrival of Governor Christopher Patten and his efforts since October 1992 to expand grassroots democracy. They formed their own political parties, including the United Democrats of Hong Kong, and supported Patten's extensive package of reforms, enacted over Beijing's strong opposition in June 1994. They also actively contested the direct elections, including those for the Legislative Council (Legco) in 1995, that were provided under Patten's reforms. Democratic activists hence have been gaining enhanced influence in Hong Kong that Beijing clearly finds unsettling.

Both Beijing's and Patten's actions have inflamed what in the mid-1980s seemed to be a stable Sino-British compromise understanding about Hong Kong's future, epitomized in the Joint Declaration of December 1984 that established the framework for the transition. First there was the Tiananmen tragedy. Then there were Patten's reforms, which Beijing saw as hypocritical, destabilizing, and directly counter to the 1984 Sino-British understanding.

Beijing made matters worse by vowing to annul the reforms, to dissolve the partially elected Legislative Council, to scrap the Hong Kong Bill of Rights, and to consider requiring reappointment of judges after 1997.[41] Many of these threats explicitly ran counter to the principle of "one country, two systems" that was enshrined in the 1984 agreement for fifty years after 1997. The ambiguous mid-1995 Sino-British compromise on Hong Kong's future Court of Final Appeal did not help matters either.[42]

China and Britain also had a running series of economic disputes over Hong Kong that paralleled the emergence of democratic reforms and that have been, in the view of many, fueled by it. Among the most conspicuous has been a three-year struggle (1991–94) over the financing of a new airport, to be completed in 1997 on Lantau Island and designed to serve Hong Kong for twenty to thirty years thereafter. Britain wanted to finance the airport as cost-effectively as possible by debt, while China accused it of inappropriately mortgaging the future of a Chinese territory-to-be with insufficient consultation.

Even after a compromise was achieved, new disputes and snubs continued to appear. At one point in late 1994, for example, Lu Ping, the director of the Chinese State Council's Hong Kong and Macao office, celebrated the inaugural issuance of Bank of China banknotes for Hong Kong, in Hong Kong. Yet Lu neglected to invite the governor, or even meet with him, during his extended visit.[43]

The 1989 Tiananmen massacre and the ensuing tensions between Britain and China have stimulated a continuing business and professional exodus of local Hong Kong residents that has contrasted sharply to international euphoria about China's own economic future. A steady fifty to seventy thousand disillusioned citizens a year have been leaving the colony, amid estimates that as many as 15–

20 percent of Hong Kong's entire population, mainly professionals, may ultimately relocate elsewhere as a result of the transition.[44] Some major businesses are also relocating; Jardine Matheson, one of the largest *hong*s, or Hong Kong expatriate trading firms, with a century and a half of residence in the colony, has reincorporated in Bermuda. Yet most of these Hong Kong émigrés, both individual and corporate, will no doubt continue to deal with Hong Kong and to influence its delicate transition, if only from afar.

As the South China Economic Zone emerges economically as a counterweight to North China, so also is an intellectual and political counterweight to Beijing emerging. Its leverage is clearly increased by the growing economic weight of nearby Taiwan and the ongoing transition in Hong Kong. Shanghai, the Szechuan area, and southern Manchuria, among other regions, all have the domestic resources and international connections to display considerable autonomy from Beijing. This proliferation of power centers within China may, over the coming decade, steadily encourage centrifugal tendencies across that huge nation as a whole.

A Complex Power Transfer Across the Generations

Compounding the problem of national unity is ongoing generational transition: the Long March generation, which masterminded the Revolution of 1949, is steadily passing from the scene. It is now over sixty years since Mao Zedong's ragged band trekked more than six thousand miles in 370 days, across dozens of rivers and mountain ranges, suffering over 90 percent attrition, en route from nearextermination in south-central China to safety in Yenan. Even the Revolution is nearly fifty years distant.

It was these epic experiences of struggle that unified party, army, and other key elements of mainland Chinese society, forging the interpersonal networks that keep that giant country unified and politically stable. The same can be said in microcosm for Taiwan, where the Civil War compatriots of Chiang Kai-shek (and opponents of Mao) are virtually all gone from the scene. But as in Beijing, no clear alternative order has yet taken their place, despite the extensive democratization of recent years.

The passing of the Civil War generation, on both sides of the Taiwan Strait, raises not only questions of individual succession, but even more important issues of institutional persistence. How will the Chinese Communist Party (and the Kuomintang on Taiwan) shape the politics of the future? And what of the military's role?

Unifying China Through Nationalism

With social tensions, regionalist pressures, and generational leadership transition all tearing at the cohesion of China, nationalism stands as an attractive way of pulling the nation back together. For a country with the largest population, the longest continuous recorded history, and arguably the most brilliant cultural heritage on earth (not to mention continuous double-digit growth for more than fifteen years), national pride comes naturally. Yet Chinese, looking back on what they view as two centuries of gross Western ethnocentrism since the Industrial Revolution, feel that they have not been given their due. Leaders making nationalist appeals have much to build upon in the Middle Kingdom.

China borders on fifteen countries—more than any other nation on earth. It defends land frontiers over seventeen thousand miles long.[45] Five percent of the entire continental shelf on earth abuts its indented eleven thousand miles of coastline. Not surprisingly, given China's imperial history and collective national hubris, painfully insulted in recent years, it has a formidable list of unsatisfied national territorial claims along its sprawling borders. These provide a fertile ground for conflict with its neighbors.

Most consequential for the future are the offshore territorial disputes, intensified by China's rapidly rising food and energy requirements. China is rapidly becoming a major oil importer, and its consumption of seafood has quadrupled since 1978.[46] In February 1992 the National People's Congress issued a "Law on the Territorial Waters and Their Contiguous Areas" that claimed 80 percent of the South China Sea for China as well. Malaysia, Brunei, the Philippines, Indonesia, and, most important, Taiwan, Japan, and Vietnam all have conflicting maritime claims somewhere in the East or South China Sea. On several occasions these have led to violence on the seas, particularly between Vietnam and China. In contrast to the United States and the former Soviet Union, China has no buffer of friendly states among its numerous neighbors. To the contrary, many are persistently hostile, encouraging a stronger tendency toward active use of military force in border disputes over the past half century than in any other major nation.[47]

On land as well as on the sea, China confronts important and delicate territorial disputes. Across the Himalayas, there is the continuing struggle with India, which erupted into violence during 1962. Nearer to China's populous and prospering southern centers there are bitter differences with Vietnam, some nearly a thousand years old. These exploded into violence during 1978.

To the north there is the tortuous, winding, many-thousand-mile frontier with

Russia, which erupted in border conflict along the Ussuri River in 1969. Many of the complex local Sino-Russian disputes have been recently settled, in protracted talks that have continued for nearly a decade. Yet the stark underlying reality of irritants created more than a century ago by "unequal imperialist treaties" still remains. Jilin province in Manchuria, for example, lacks access to a Sea of Japan that is visibly within sight of a major Chinese industrial center at Hunchun, because of the Treaty of Beijing, imposed on a weak and divided Manchu government by czarist Russia in 1860.

China, as *New York Times* correspondent Nicholas Kristof points out, shares with the imperial Germany of 1900 "the sense of wounded pride and the annoyance of a giant that has been battered and cheated by the rest of the world."[48] Not surprisingly, given this historical background and its new energy vulnerabilities as a rising oil importer, China has used a substantial fraction of the dividends from its recent explosive economic growth to embark upon a sustained and major military modernization program. Domestic political imperatives to unify the nation through nationalist symbols and to stabilize the leadership's political base naturally give this military "enchantment" further momentum.

Deepening Defense-Industrial Relations with Russia and the Islamic World

The collapse of the Soviet Union has afforded China an enormous new opportunity to enhance its military and military-industrial base. Huge amounts of hardware and technology, especially in the aerospace and nuclear fields, that would have taken China years to develop can now be had very cheaply from the Russians. Hundreds of Russian and Ukrainian military scientists are also working in Chinese defense industries, with others in intimate communication through E-mail and other electronic systems. Most are pursuing the development of armaments and weapons technology banned to China by other advanced nations.[49]

Through cooperation with Russia, leveraged by domestic efforts, China is developing the clear capacity to project decisive power into neighboring areas, including the deepest reaches of the South China Sea. It has negotiated with Russia to buy up to seventy-two advanced Su-27 fighters capable of providing close air support for naval operations at extended range. Indeed, for limited periods these planes could provide air cover for naval task forces throughout virtually all of the South China Sea.[50] China has also acquired critically important in-flight refueling technology for its planes, developed mobile conventional-warhead missiles to threaten any foe within a thousand miles, and organized an amphibious rapid deployment force based on Hainan Island, ready for operations in the South China Sea.[51]

The China Seas and the sea-lanes to the Gulf are simultaneously becoming more important. China is also moving steadily toward blue-water naval capability. By around the year 2006, its two prospective forty- to fifty-thousand-ton aircraft carriers, together with related carrier groups of destroyers, frigates, and related missile systems, are expected to be operational.[52] Although important technical problems remain with respect to steam catapult systems for launching aircraft, as well as the carrier aircraft themselves, China has been actively training carrier pilots at the Guangzhou military academy. It has also been developing advanced antiaircraft radar and antisubmarine sonar systems that will be crucial in defending its new carriers.[53]

China's current operational capabilities in the South China Sea are particularly threatening to neighboring Vietnam, with which China clashed in 1974, 1978, and 1988 on territorial questions, and to the Philippines, which has recently had a conspicuously weak and poorly equipped military.[54] It appears to be no accident that China chose oil-rich waters close to the Philippines for a 1995 assertion of sovereignty, although this provoked a major Philippine buildup in response. China's limited antiair capabilities and still limited aerial refueling capacity leave its forces temporarily susceptible to air and surface attacks by other countries equipped with or acquiring advanced fighters like the F-16 and the MiG-29, or Exocet antiship missiles.[55] But this strategic calculus could change as China's blue-water navy develops. Its shadow over the China Seas will doubtless deepen toward the end of this decade and beyond.

Apart from regional conventional capabilities, the Chinese military is also making major strides in the strategic-nuclear area. Since the late 1980s, as noted in Chapter 4, China has had operational nuclear strategic ballistic missile submarines (SSBN) of the Xia class, capable of potentially striking the United States,[56] together with at least four land-based ICBMs capable of reaching North America. Since 1992, China has reportedly been perfecting MIRVed strategic missile weaponry, in an ongoing series of nuclear tests that make it, together with France, one of the few nations in the world actively testing nuclear weapons.[57] Senior Australian defense sources suggest that within fifteen years, China will have fifty to seventy ICBMs in mobile launchers, many of them MIRVed, with a range of up to twelve thousand kilometers.[58]

To be sure, the Chinese military has major technological defects in some areas. Apart from its problems in developing aircraft carriers, it lacks supersonic fighters with all-weather air-to-air combat abilities, over-the-horizon radar, and advanced submarine capability, for example.[59] But the current trajectory of its modernization is rapid, and the political incentives of a unifying nationalism for further improvement are firmly in place.

In both the conventional and nuclear areas, a crucial issue for the future security of East Asia and the world is how extensively Chinese nationalism comes to express itself through the "Islamic-Confucian connection."[60] As China becomes a major oil importer, its ties with the low-cost global oil producer, the Middle East, will naturally deepen. How will they evolve in the security dimension, particularly with major oil exporters such as Iran and Iraq that want sophisticated arms and technology and are, like China to some degree, on distant or estranged terms with the West?

Beijing has had a substantial arms market in the Middle East for roughly the past fifteen years, although its arms sales have declined from a high point during the Iran-Iraq war of the 1980s. One of the major early coups was a $3 billion 1985 arms deal between China and Saudi Arabia for CSS-2 missiles, after the U.S. Congress had denied Riyadh's request for F-15E fighter-bombers that it feared might compromise Israel's security. The Saudi tie, however, has since atrophied.

Around 95 percent of Chinese arms exports are delivered either to countries within 150 miles of China's borders or to the Middle East.[61] Iran, Iraq, and Pakistan have recently been the primary Middle Eastern recipients, purchasing large quantities of tanks, ships, submarines, and especially missiles from China. Following the large 1992 U.S. sale of advanced fighters to Taiwan, for example, China's leaders quickly approved a reciprocal sale of nuclear-capable M-11 missiles to Pakistan, where they currently deepen the accelerating arms race on the Indian subcontinent.

Among more than $1 billion in Chinese weapons sales to Iran since 1989 have been the HY-2 Silkworm and the C-801 antiship missiles. China has also set up an assembly plant in Iran to produce both the M-9 and the M-11 intermediate-range ballistic missiles.[62] During the mid-1990s, Iran also became the first nation to receive China's advanced CSS-8 short-range surface-to-surface missile, with a range of about one hundred miles.[63]

Among the most vexing questions for the future is how much nuclear technology China will ultimately supply to Iran and Iraq.[64] In 1990, China and Iran concluded a ten-year nuclear cooperation agreement. Beijing agreed not only to build research reactors and to supply calutrons, which play a key technical role in uranium enrichment, but also to teach the Iranians how to operate all this sensitive equipment.[65] Tehran has since spent millions of dollars on Chinese equipment for producing highly enriched uranium, an essential component of nuclear weapons, while also aiding the building of several Iranian nuclear plants.[66] Two of these have been installed at Iran's Isfahan center for nuclear research, with its growing nuclear cadre of more than three thousand personnel.[67]

Ultimately it is the combination of Chinese nuclear and missile-related assistance to Iran, supplemented by related aid from Russia and North Korea, that is most destabilizing. Emerging Iranian capabilities threaten not just surrounding Arab states like Iraq, Kuwait, and Saudi Arabia, but Israel as well. Chinese Silkworm missiles, if deployed by Iran above the narrow Strait of Hormuz, would also lie in watch over the strategic entrance to the Persian Gulf, through which passes half of the oil trade on earth.

Fragile Interdependence with the Broader World

Despite rising capabilities that could lead to more militant, nationalistic power projection, China most likely will be constrained in its militancy by deep—and still rising—economic interdependence with the world, especially the major advanced industrial nations. The share of trade in China's economy doubled overall over the 1980s, with trade as a proportion of GNP rising to 18 percent by 1993.[68] By 1994, China's overall foreign trade had grown to $235 billion, up from just $70 billion in 1985.[69] And the prospects are for still further expanded trade and related economic interdependence with the outside world in future.

Apart from finding markets for manufactured exports and sources of energy imports, the simple arithmetic of feeding the huge population living within its borders will deepen China's stakes in trade interdependence. With a fifth of the world's population, China has only a fifteenth of its arable land. With rural modernization, grain production is land- rather than labor-intensive. Theories of comparative advantage dictate that China should move away from commodity farming and into manufacturing, which seems to be exactly what is happening.

A rising taste for meat in China's diet intensifies the demand for food imports, especially feed grains, still further. It takes Chinese farmers at least four pounds of feed grains to render one pound of meat.[70] And imported customs have stimulated very sharp increases in that meat demand, especially for lesser-eaten varieties such as beef. Hamburgers sold by the four McDonald's franchises in Beijing alone pushed up the city's beef demand by 150 percent in one year.[71]

In 1994, China's grain crop slipped 2.5 percent to 444 million tons, as arable land declined by 2.7 million acres.[72] Purchases of foreign grains shot up 34 percent. In 1995 they doubled to around fifteen million tons a year. Recent forecasts suggest that Chinese grain imports may well reach fifty million metric tons early in the next century. By 2030 they could reach ninety million tons, or roughly half of total current global grain exports.[73]

In some of China's largest and most affluent provinces, trade dependence has already reached levels rivaling those in capitalist Asia. In Guangdong, for example, exports are nearly 20 percent of GNP, compared to only 9 percent in Japan, and 7 percent in the United States.[74] Hainan's 15 percent, Shanghai's 13 percent, and the 12 percent of both Fujian and Liaoning are also close to international levels. Trade-dependent Guangdong, Shanghai, and Liaoning are all among the small number of areas gaining enhanced Politburo representation in 1993, as noted earlier. The profile of internationally oriented Shanghai looms especially large in the political economy of the post–Deng Xiaoping era.

Even more important to China's future than trade interdependence with the world will be capital flows. Aided by massive exports fueled by low-cost labor and heavy foreign investment, China doubled its foreign exchange reserves to nearly $50 billion in 1994.[75] But like a growing teenager, consuming ravenously in order to grow, the Chinese economy has huge borrowing needs for the future.

For these needs to be met, one would assume that China would seek to pursue a moderate course in international affairs. Yet there is no assurance that the domestic pressures outlined above might not force it toward a more militant, nationalistic line, especially in an era of political transition such as that now impending. Indeed, there were signs of such a tendency in sudden confrontations with Taiwan, the Philippines, and Vietnam, among others, during the mid-1990s. Certainly erratic oscillation between chauvinism and internationalism has been a hallmark of China's recent past, and it may well dominate the foreseeable future.

To an unusual degree for a major nation on the global scene, even China's near-term prospects are clouded in uncertainty. But the broad parameters of the long-term future and the dimension of the challenge that this enigma may someday pose for the world are certainly clear. China has over four times the population of any other major developed nation on earth. Its economy has quadrupled in size since 1979, to levels probably exceeding a tenth of the entire world's output. Yet Chinese per capita income is still low, and the prognosis for rapid future growth is high, for at least the next two decades, as China closes the gap between its own human living standards and those of the broader world. An enigmatic China, historically prone to shift its vast weight in the global scales of power as it seeks advantage, will no doubt cast a larger shadow over Asia and the world in the next generation than it does today.

ASIA'S NEW BALANCE-OF-POWER GAME

As the Cold War wanes, the shadow of China lengthens, and Japan's offshore economic stakes grow ever stronger, international relations within Asia begin to take on a fluid new cast, subtly changing the challenge of Pacific Defense. No longer do superpowers hold sway and dictate to unequal, and thus subservient, junior allies. Earlier sections of this book, particularly Chapter 2, have shown us the delicacy of domestic political circumstances across Asia, especially in the Northeast Asian Arc of Crisis surrounding Japan. This chapter focuses instead on how those precariously balanced political systems, many in the throes of historic generational and institutional change, actually interact with one another. The Asia of the future, we shall argue, poses not the prospect of a reemerging Japanese coprosperity sphere but a different, and potentially more unsettling, danger with global implications: the specter of Europe's bloody and turbulent balance-of-power past.

Why should Asian instabilities and regional rivalries be a threat to America's own Pacific Defense? They can, of course, directly threaten American economic interests, including trade and investment by American firms in the Far East. More important in the long run, they could lead to fundamental reorientations adverse to American interests, and destabilizing in global terms, in major nations of the region, particularly Japan.

CHAPTER 7

Some see Asia's new balance-of-power game as a golden opportunity for the United States to play Asian nations against one another, and to elicit new benefits for the United States.[1] This is a dangerously shortsighted view. As the following pages will suggest, Asia's new balance-of-power game has the explosive potential—given underlying resource rivalries and the dual shadows of Chinese and Korean reunification—to generate destructive, wasteful regional arms races, and possibly to stimulate Japanese rearmament. The imperative for America and for a farsighted Pacific Defense is dampening regional rivalries and uncertainties rather than inciting them.

For nearly three centuries from the mid-seventeenth century until World War II, the Prussians, Russians, French, Austrians, and English, in various incarnations, schemed in shifting coalitions for political advantage that none could ever sustain. The result was a series of wars, many in miscalculation, that ended in two global conflicts in this century that together killed nearly 100 million people. Only with the emergence of the Cold War, ironically, did Europe attain a degree of peace. And even that proved a politically destructive fate, consigning the proud Continent to but a peripheral ensuing role in global affairs.

Richard Nixon and Henry Kissinger brusquely decreed the birth of balance-of-power politics in Asia twenty-five years ago, as they maneuvered between the Communist Sino-Soviet giants and administered "Nixon shocks" to their erstwhile allies in Tokyo. But even the modest success of their diplomacy, ironically, came only because a true balance-of-power setting, in the nineteenth-century European sense, did not exist yet in Asia. They could be cavalier and unpredictable with their Japanese, Korean, Taiwanese, and Vietnamese allies precisely because those allies had nowhere else to go. The Asians were all dependent on Washington, in both security and economic terms, but enjoyed few mutual links to one another.

Today there is emerging a very different, and more fluid, new world that is coming to manifest many of the classic European traits.[2] The key nations of Asia are generally far less dependent on the United States for trade and investment than was true two decades ago, and conversely more interrelated to one another. These nations are also substantially stronger relative to America in economic, technological, and even to some degree military terms. Both the United States and its longtime allies are also freer to disagree than previously, with the political solidarity imperatives of the Cold War at least temporarily relaxed. Thus any significant increases in strength or diplomatic advance by one of them is increasingly likely to provoke fluid offsetting behavior by others, as was so common in Europe a century ago.

There is one crucial caveat that must be made in applying European analogies to East Asia. Value systems and concepts of strategy are in many ways significantly different. Western values of the past three centuries have tended much more toward absolutism than those of East Asia, whose experience has no analogues to the Crusades, the Thirty Years' War, or the Inquisition. East Asian pragmatism may moderate the force of regional conflicts.

So too may East Asian concepts of strategy. They place strong emphasis on indirection and psychological pressure, avoiding violent techniques except as a last resort. Sun Tzu's dictates that "victorious warriors win first and then go to war" and that "to win without fighting is best" express the outlook with which much of Asia approaches balance-of-power struggle.[3] Yet to highlight such psychology is not to suggest it is always determining. The uncertainties and dangers of miscalculation in a balance-of-power world are innate, and to a significant degree universal.

The New World of Seven

In Asia's new political environment, the seven distinct poles of an emerging Asian power game are growing ever more clear. Apart from the Cold War superpowers (the United States and Russia), together with an emerging superpower (China) and an economic giant (Japan) that could vie for such a role, there are several regional powers of significant economic and political promise. India has nearly a billion people, undeclared military nuclear capacity, and major unrealized economic potential. Vietnam has one of the largest and toughest conventional armies in the world, and its economy is growing explosively as the constraints of a long, post-1975 isolation come to an end. Korea, although still divided, is moving toward probable union, when the combined mass of its northern and southern halves will be internationally substantial, in military, economic, and technological terms.

As indicated in Table 7-1, each of these key Asian players varies in its dimensions of strength. But most have at least 100 million people, nuclear or near-nuclear capacity, GNPs above $300 billion, and at least half a million troops under arms. The region clearly has, in the post–Cold War dawn, multiple centers of prospective political, military, and economic influence.

Apart from the seven key players noted above, at least four others—Australia, Indonesia, Taiwan, and Malaysia—also seem destined for highly strategic, albeit more specialized, roles in Asia's emerging power game. Despite its small population of around twenty million people, Australia has potential to play a pivotal mediating role among larger powers in the region, in support of stable regional security and economic frameworks. In support of this role, it maintains a sizable defense budget (nearly the size of India's), outstanding intelligence capabilities,

Table 7-1

PROSPECTIVE KEY PLAYERS IN THE EMERGING ASIAN POWER GAME

Country	GNP ($ billion)	Population (million)	Defense Budget ($ billion)	Armed Forces	Nuclear Capacity
Japan	4,592	125	53.8	239,500	near nuclear
China	509	1,201	7.5	2,930,000	yes
Korea (North and South)	401	69	16.6	1,761,000	residual capacity
Russia	1,120	149	62.8	1,520,000	yes
India	270	934	8.1	1,145,000	undeclared
Vietnam	19	74	0.9	572,000	no
U.S.	6,737	263	252.6	1,547,000	undeclared

Source: IISS, *The Military Balance*, 1995–96 ed.

Notes: All figures for 1994, except defense budgets, which are for fiscal 1995. Actual Chinese defense expenditures are believed to be much higher than the nominal defense budget, mainly because of the use of profits by defense enterprises for military purposes. Estimates for the 1995 defense budget, for example, range from $7.5 billion to $50 billion. See *Far Eastern Economic Review*, April 13, 1995, p. 25.

and a formidable, well-maintained network of human ties spanning East and West.

Taiwan's importance in some ways flows from parallel international assets to those of Australia, despite its government's difficult diplomatic position as the loser in the 1948–49 Chinese civil war. Taiwan is a small island with Australia's limited population of around twenty million people. Yet it maintains outstanding intelligence and informal diplomatic networks throughout the Pacific, leveraged by the pervasive role of overseas Chinese in political economies across the region. With a defense budget well over $10 billion annually and nearly half a million men under arms, Taiwan has substantial military capabilities. Its position may also be strengthened in future by rising anxieties within the region about Beijing, to the extent that it remains in tension with the mainland. This balance-of-power dynamic among China, Taiwan, and other nations of Asia is already apparent in Vietnam. There informal economic and even defense-industrial ties with Taiwan are deepening rapidly, despite a history of intense mutual antagonism between North Vietnam and Taiwan during the Cold War.

Indonesia also figures in East Asia's power game, and should do so more significantly in future. The best comparison is with India. Although smaller and lacking in nuclear capacity, Indonesia has many of India's other geostrategic assets, and it lies across key sea routes that will become the conduit for rising Asian energy imports from the Middle East. Its ability to exercise leverage on these routes is rising, with its rapidly increasing air and naval expenditures.

Last but not necessarily least, there is Malaysia. Like both Australia and Taiwan, it has a small population of only twenty million, but excellent international networks, especially in Northeast Asia, which its top leadership informally visits every few months. Its efficient, rapidly growing economy is highly regarded by both U.S. and Japanese multinationals as a manufacturing base; with around $50 billion a year in exports, it outpaces both Australia and Indonesia. Malaysia's veteran leader Mohammed Mahathir has long had a strong personal inclination to play balance-of-power politics. His ability to do so in the mid-1990s was further enhanced by his strong advocacy of trade regionalism, rapid expansion of defense spending, and decision to source military equipment from both Western nations and Russia.

This emerging pattern of multiple, relatively balanced powers in Asia is a sharp departure from the recent past. For the past half century, of course, the America that won the Pacific War in 1945 has loomed much larger than any of the others. Yet recent shifts in economic power and political inclination, stimulated further by regionalism, have altered this picture significantly.

New Actors and Intensified Uncertainties

The emerging East Asian power game seems likely to feature several national players who either did not exist or did not figure prominently on regional questions in the past. Most conspicuously, there is Korea. A unified Korea, of course, does not yet exist, and will most likely emerge only gradually—if possibly in rather turbulent fashion—over the coming decade. But its birth seems virtually inevitable. There are huge economic differentials: poverty-stricken North Korea now has only one-tenth the GNP of the affluent South. And Seoul's ties with Pyongyang's major erstwhile allies in Beijing and Moscow are rapidly flowering.

North Korea will gradually become a de facto client of the colossus to the south, and unity, through de facto collapse and absorption, will gradually but inevitably arrive. This development will crucially shift power configurations in Northeast Asia and serve as an important catalyst to more active, and probably more conflictual, balance-of-power diplomacy in the region. Some of the possibilities are already clear in subtle jousting among China, Russia, and Japan, provoked by Seoul's activist Northern diplomacy since 1990.[4]

When it does emerge, the new unified Korea will be a major force in Northeast Asian affairs, not least in military terms. North and South Korea together, as Table 7-1 suggests, have more men under arms than either of the superpowers, the United States and Russia. Clearly a unified Korea could not maintain this degree of mobilization without greatly alarming its neighbors, and it would find the economic costs taxing in any event. But even a combined North-South force of one-third current levels would be formidable. A unified Korea would also have a population 90 percent the size of Germany's and a GNP approaching that of Brazil, even before the sustained growth likely to ensue as the impoverished North moves ever closer to South Korean living standards.

Korea's movement toward reunification threatens to pose complex new challenges for Japan–Korea relations—and for the U.S.–Japan–Korea strategic triangle—that are insufficiently perceived at present. Japan–South Korea relations are already tense; as a recent joint bilateral opinion poll pointed out, 68 percent of Koreans actively dislike Japan, and Korea is one of the most unpopular countries in Japan as well.[5] More Japanese, in contrast, doubt that Korean reunification will occur, and could well feel threatened by the prospect, which this analysis suggests is quite possible in the coming decade.

Korean reunification could also fan nationalist sentiments within Korea, which recent polls suggest are quite strong and anti-American.[6] It could therefore accelerate pressures for an American military withdrawal from Korea, which could in turn intensify parallel pressures within Japan for an American withdrawal. A combined withdrawal of U.S. forces from both Korea and Japan would also probably intensify pressures for expanded military spending in Japan, and the dangers of a Sino-Japanese arms race, to be considered later in this chapter.

Vietnam, like Korea, has never been a functioning, integral member of the modern East Asian international system. But the latent political economic implications of Hanoi's 1975 victory in the Vietnam War are at last being realized. With over seventy million industrious people, a battle-hardened military of more than half a million men, and extensive resources, a unified Vietnam is a natural power in Southeast Asia. With membership in ASEAN and deepening relations with the United States, Japan, and Australia, together with explosive economic domestic growth, its long-standing potential is now being steadily transformed into reality.

India, although territorially cohesive since the partition of 1947, has not traditionally been a major figure in East Asian international relations, despite participation on the Indochina armistice commission for twenty years (1955–75). But

New Delhi is nevertheless likely to figure with increasing regional prominence in East Asia, for four reasons. First, its overall geostrategic role and power-projection capabilities have steadily risen over the past two decades. Since 1974, India has not only become a de facto nuclear power but has also purchased two small aircraft carriers and developed sophisticated missile delivery systems. In addition, Asia's energy dependence on the Middle East—and especially that of India's traditional adversary China—is rapidly deepening. This means that the sea-lanes from East Asia to the Gulf, together with such intermediate states along that path as Myanmar, are assuming increasing strategic importance. Many of those areas are also within India's clear sphere of geographic concern. India's deepening economic ties with Asia, stimulated by that region's explosive economic growth, and its reciprocal interest in India's potential will also help draw India politically to the East. Rivalries with a dynamic and growing China could have a parallel effect.

Why Balance-of-Power Rivalry?

Five to seven is a classic number of prospective players for vigorous balance-of-power politics. Indeed, the structure of East Asia's emerging power game strikingly resembles, in that respect, the tragic eighteenth- and nineteenth-century configurations in Europe that provoked the Seven Years' War, the Napoleonic Wars, World War I, and World War II. But why should a mere structural similarity in the number of players reproduce in East Asia the tragic outcomes of Europe's past?[7]

Three basic factors are at work. First, there is the character of the nations in East Asia itself. Second, there is the nature of their interdependencies, or the lack thereof. Finally, there are the prospective costs and benefits of conflict, insofar as we can forecast them within the East Asian power game that is now emerging. Signals in all three areas point to serious danger ahead.

As in the Europe of Nicholas II, Kaiser Wilhelm, and David Lloyd George, but not the Western Europe of the recent past, there is sharp diversity among the political systems of Asia today. At one end of the continuum, Japan is a democracy with competitive elections on the Western model, albeit a democracy with strongly communitarian features. Yet China, North Korea, and Vietnam, in particular, where Communism has persisted into the post–Cold War world, are organized very differently. Russia, South Korea, and Indonesia are other variants still. Democracies may not fight one another, but mixed collections of hybrid democracies and dictatorships, like those found in East Asia today, very often do.

An era of rapid domestic growth and industrialization is often also one of external expansion. Samuel Huntington has pointed out, for example, that the external expansion of the United States, the Soviet Union, Germany, Japan, the United Kingdom, and France, in the nineteenth and twentieth centuries all coincided with periods of intense domestic industrialization and economic development.[8] The major nations of Asia, especially in its northeast segment, have the similar economically rooted hubris of emerging powers.

Large, emerging powers come naturally to balance-of-power tactics, as the past diplomacy of Bismarck and Teddy Roosevelt suggests. Such nations have natural economic and geostrategic leverage because of the mingled fear and anticipation of others regarding their future potential. Such states are also often less embedded in established relationships than other powers and thus correspondingly free to shift sides, in balancing fashion, as national interest dictates.

China is a clear case of a nation with strong incentives to play balance-of-power politics. It has the leverage of a large, rising power and the detachment of one without established allies. The immense cultural self-confidence of a brilliant four-thousand-year history and a distinctive revolutionary tradition reinforce this aloofness from entangling alliances. Not surprisingly, China since at least 1949 has manifested a clear tendency in its diplomacy to actively pursue balance-of-power politics.[9] This tendency, if anything, has been strengthened by China's emerging international economic prominence in the post–Cold War world.

Like China, Russia has also traditionally played balance-of-power politics in the East Asian region. As a large outsider in the established diplomacy of the region, with massive resources and apparent potential as a market, it has had resources to confer on prospective collaborators. In the background, its massive military has also provided leverage.

In recent years, South Korea has also become a convert to the balance-of-power game. With the waning of the Cold War it began to develop large and potentially lucrative economic ties with both China and Russia, outflanking the North. It also began subtly using these ties with the continental Communist giants to pressure Japan. With a strong economy, substantial military, cohesive decision-making apparatus, and central location in Northeast Asia, Seoul is well equipped to manipulate rivalries among larger powers of the region to its own advantage, and it has actively begun to do so.

Smaller nations of the region have also found incentives to manipulate the emerging regional balance of power, finding their maneuvering ability a valuable asset. Taiwan, for example, has consorted with Russia, China, the United States, and Vietnam. The Malaysia of Mohammed Mahathir has linked itself variously to

Japan, Korea, Russia, and Vietnam, in an attempt to offset American influence. Vietnam itself has similarly deepened ties with Japan, the United States, and Malaysia, to offset its powerful neighbor China.

The one nation so far finding the balance-of-power game uncongenial has been Japan. For both cultural and structural reasons the complex, consensus-oriented Japanese foreign policy apparatus takes time to arrive at clear decisions. It has found it hard and distasteful to maneuver among the nations of the Pacific, and has preferred, for the moment, to continue in its traditional path of "reactive" diplomacy responsive to American initiative.[10] While this passive stance, by no means ideal, generates periodic frustration in the United States—so clearly evident during and just after the Gulf War, despite Japan's $13 billion contribution to the war effort—it is far preferable, from an American standpoint, to the Gaullist, more fully rearmed Japan that could emerge from an American immersion in the manipulative balance-of-power politics favored by China, Malaysia, and others.

Looking to the future, as the economic and political shadow of balance-of-power advocates, preeminently China, rises across the region, their diplomatic proclivities will likely exert more and more influence on the international relations of the region as a whole, in the absence of stabilizing American efforts. If China manipulates access to its market in balance-of-power fashion, for example, other nations may have little choice but to act likewise. Existing Asian "hub and spokes" alliance relationships with Washington could come under substantial pressure as the nations of the Pacific are increasingly encouraged to balance among Washington, Beijing, and other major capitals by a combination of new economic and political imperatives.

Balance-of-power politics obviously requires not only nations with incentives to engage in it, as opposed to more prosaic alliance diplomacy. It also requires a catalyst for balance-of-power-oriented struggle. As earlier chapters suggested, there are many such catalysts across the region. A large and surprising number of them concern energy, Asia's Achilles' heel. After all, China, Japan, Korea, and ASEAN are all emerging as vigorous prospective buyers in global oil markets, as noted in Chapter 3. Many tensions also flow from territorial disputes, especially with China, and related emerging struggles over marine resources.

The clearest and most obvious point of contention—for literally all of Asia's major local powers—is the South China Sea. It is both an "energy" and a "territory" problem simultaneously—a volatile combination in the energy-short world toward which we may now be moving. Beneath that sea's placid, tropical aquamarine surface may well lie substantial reservoirs of oil, which all the powers of the region covet. Through its waters run shipping lanes to and from the Persian

Gulf that carry 80 percent and more of Northeast Asia's precious oil supply. And China also claims 80 percent of the sea's entire area, in assertions disputed by five other nations.

Conflicting maritime claims over the continental shelf to the north, and its natural resources, are an even more dangerous prospective catalyst for balance-of-power conflict. The oil reserves of the East China Sea between Shanghai and Okinawa, in particular, look tempting. And they are bitterly contested, among multiple powerful claimants.

More than 5 percent of the world's entire maritime continental shelf is adjacent to China, in a broad plateau sweeping outward from its eastern coasts for nearly a thousand miles in some places. Naturally, the resources of this shelf become a bone of contention within East Asia. But apart from this, territorial claims are a catalyst for conflict all across the region, especially on the fringes of China.

India and China disagree over their Himalayan border, and fought a bitter war to define it in 1962. Vietnam and China have border conflicts, on both sea and land, that led to open violence in 1974, 1978, and 1988. Russia and China disagree also, and came to blows in 1969. Although many border irritants have since been resolved, the residue of the unequal 1860 Treaty of Beijing, which deprived Manchuria of most of its sea access, still remains, to defy the political-economic logic, not to mention the nationalist claims of a growing and ever more assertive China.

In the combustible, dynamic world of East Asia's new power game, the links that bind its players and constrain their strivings are sadly and dangerously skewed. To be sure, there is an informal, person-to-person dimension to integration, especially within Greater China and between Japan and parts of Southeast Asia (in particular Malaysia, Indonesia, and Thailand). This Westerners often miss. There are also important (if less elaborate) back channels between Japan, China, and North Korea. Chōsen Sōren (the association of ethnic Korean sympathizers of the North in Japan), for example, has deep ties with both the Japanese Social Democratic Party and North Korean leadership. But conspicuous gaps in cross-regional networks and in cross-regional political-economic relations clearly exist to a greater degree than in Europe or North America. Regional political unity that transcends both nation-states and their balance-of-power propensities will be difficult to achieve.

The gaps in both cross-regional economic interdependence and human networks are most striking—and most serious—among the big powers of the region. India and Vietnam, for example, are both distant and estranged from China. Economic and political distance is also dangerously common among large states in

Northeast Asia. Japan, in particular, is growing relatively more distant from the Northeast Asian continent,[11] where economic and human linkages among North China, South Korea, and the Russian Far East are deepening much faster than those with Japan. To be sure, there are exceptions to this pattern of estrangement, such as the explosive growth of Japanese investment in Dalian, at the southern tip of Manchuria. Yet they are only exceptions to a broader pattern of growing relative Japanese isolation.

Informal trading networks, to be sure, play enormously important, and generally effective, roles in coordinating economic transactions across Asia, even without customs unions or other formal institutions of integration. The Japanese *keiretsu*, the Korean *chaebol*, and the overseas Chinese family businesses are all structurally diverse, but functionally parallel, cases in point. Yet whatever their agility in the economic sphere, such groups are simply incapable of providing the security framework so prospectively important in moderating future Asian balance-of-power rivalries.

Asia, it is fair to say, is "strikingly underinstitutionalized" compared to Europe or North America[12]—a reality that is both a cause of disturbing balance-of-power tendencies and a reflection of them. It clearly has no equivalent to either NATO or the European Union. SEATO, which John Foster Dulles hoped to see evolve into an Eastern NATO, was defunct within a decade of its foundation in the mid-1950s. ASEAN is still an anemic substitute, even with the recent addition of Vietnam. APEC in the economic sphere and ARF (Asian Regional Forum) in security affairs are both embryonic and find it difficult to overcome preexisting national interests and loyalties.

A final factor destabilizing Asia's new multipolar power game is the structure of prevailing costs and benefits. Expansionist, confrontationist strategies, not to mention the acquisition of nuclear weapons, offer some attractive prospects of gain to regional powers, such as preferential access to energy reserves and sea-lanes in the South China Sea. The costs of armament and preparation for war, conversely, become less onerous as East Asia grows increasingly affluent. This combination of wealth and bellicosity is a recipe for disaster.

Nuclear proliferation, in particular, threatens to introduce the complexity and danger of multilateral security calculations into East Asia, increasing the risks of miscalculation in the dangerous multipolar power game now emerging in the region. Virtually all the major players, because of the region's general technical sophistication and the prevalence of plutonium-generating civilian nuclear power programs, are close to nuclear status if they do not have it already. The uncertainties of the Asian balance-of-power game now unfolding could ultimately push several nations across that ominous threshold.

Sino-Japanese Relations: Crux of the Emerging Power Game

In the emerging East Asian power game the protagonists will doubtless be the two giants of the region, Japan and China. There is, of course, a fair degree of interdependence rapidly growing between them already, with over $6 billion in Japanese investment in China and more than $40 billion in annual trade. Their cultural affinity is stirring, especially to the Japanese, most of whom can recite by heart T'ang poems learned at school. But the Sino-Japanese relationship builds on a troubled past, including memories of twenty million Chinese deaths, civilian and military, at Tokyo's hands in World War II.[13] And in the balance-of-power world now emerging in Asia, there will surely be much to provoke these latent Sino-Japanese tensions.

Energy, as we have discovered so often with respect to other East Asian problems, could be a catalyst for conflict. Japan has long been a huge energy importer. And China, despite its large domestic reserves, is rapidly moving in that direction. A generation hence, each should command roughly a third of Asia's oil imports. In tight global and regional energy markets, the two could easily become active and determined competitors.

Sino-Japanese energy tensions could be aggravated by the simple geographical propinquity of the two giants. Although they lack common land borders, China and Japan lie next to each other in the East China Sea, with Shanghai less than five hundred miles from Okinawa. Both nations claim a bevy of island shoals (known to Japan as the Senkaku Islands and to China as Diauyutai), roughly halfway between China's largest city and Japan's greatest military base, beneath which are reputed to lie huge reserves of oil. The two nations came close to violence over these shoals in 1974, and the conflict that boiled up then remains unresolved.

Energy shortages will, as noted above, most likely make both China and Japan increasingly anxious about secure access to Middle Eastern oil in coming decades as both nations' import requirements from that low-cost source begin to mount. Such energy security concerns will doubtless also prompt a rising anxiety about the sea-lanes over which ever more vital energy imports must travel. In the short run, this could produce Sino-Japanese tensions over the South China Sea, as noted above, and intensified Chinese overtures to Myanmar.[14] In the longer run, it could possibly provoke a naval arms race between the two giants, beginning with helicopter and smaller aircraft carriers and leading ultimately to competitive blue-water navies for both.

Apart from the problem of energy and its tangible expression in the deceptively placid waters of the South China Sea, Japan's and China's traditional, overlapping geostrategic concern for neighboring Korea and Taiwan could also be a fertile field for Sino-Japanese conflict. Korea has long been seen in Tokyo, as Itō Hirobumi put it nearly a century ago, as a "dagger pointing at the heart of Japan." Taiwan lies, even more decisively than the South China Sea, directly across strategic Japanese sea routes. Both areas also have major industrial potential. And both have historically been traditional parts of Greater China.

The very ambiguity of Korea's and Taiwan's long-term orientation helps intensify their potential as a flashpoint for Sino-Japanese conflict. Throughout classical history both remained at the very fringe of the Chinese Empire, yet loosely within it. Taiwan was known by the Chinese, and has been considered part of their empire since the seventh century. Yet it was long a haven for pirates, with little serious Chinese settlement until a thousand years later. The site of Portuguese, Dutch, and renegade Ming outposts, Taiwan did not conclusively yield to Chinese central authority until 1683. The Shilla dynasty of Korea forged a closely allied relationship with China against northern barbarians, and the deeply Confucian Yi dynasty, founded in 1392, paid systematic tribute to China. Yet Korea nevertheless retained an autonomous identity until modern times.

Despite their venerable ties to Beijing, neither Korea nor Taiwan have been decisively aligned with China for very long at any point since Japan first emerged as a modern state late in the nineteenth century. For nearly half a century, Japan claimed both as colonies. And since 1945 and 1949 respectively, South Korea and Taiwan have been sheltered under an American security umbrella, together with Japan, that effectively neutralizes Chinese influence. Should that situation change in the context of rising Chinese economic and military power, coupled with a declining American presence in the western Pacific, Sino-Japanese tensions could be dangerously intensified.

A fluid, multipolar East Asian power game would give China and Japan ample field for political maneuver. Japan would no doubt feel a natural community of interest with the Malay states of Southeast Asia, such as Malaysia and Indonesia, together with Thailand. In these three nations alone it has huge investment stakes totaling well over $30 billion. Major shares of Japanese oil, aluminum, textile, and consumer electronics imports flow from the diversified, Japanese-affiliated production bases in these countries. China, conversely, could make common cause with Myanmar, as well as with many of its twenty million overseas Chinese brethren. Vietnam would deepen ties with Japan and ASEAN, while strengthening relations, ironically, with its former enemy, the United States. Korea, India, and Russia could play ambiguous but pivotal roles in this volatile, wary waltz of shifting coalitions.

The Drift Toward "Arms Race Asia"

The uncertainties of the emerging East Asian power game, aggravated by energy shortage, have accelerated the dangers of a post–Cold War Asian arms race. To be sure, there are some important, built-in elements of restraint, such as skeptical ministries of finance and an "economics-first" orientation, that explosive areas like the Middle East and the Balkans do not share. Yet as indicated in Table 7-2, every one of the major nations of East Asia except Vietnam and North Korea, which already have two of the most formidable arsenals on earth, increased actual defense spending by more than 25 percent during the 1990–94 period.[15] Indeed, most Asian nations boosted military spending by nearly double that ratio. This intense arms competition in Asia has been continuing for nearly a decade, and it must be an unsettling element of the region's prospective future, in light of the political rivalries discussed above.

Within East Asia itself, the trends are especially unsettling in the northeastern part of the region, surrounding Japan in a great Arc of Crisis from southwest to northeast. As also indicated in Table 7-2, Japan, China, Taiwan, and South Korea combined probably spent nearly $90 billion on defense in 1995, even by highly conservative estimates; actual expenditures could be as much as triple these figures in the case of China, since much defense spending is reportedly camouflaged in civilian budget items and cross-subsidies from the profitable nonmilitary operations of defense enterprises, ranging from nightclubs to food production. This figure represented an increase of over 60 percent for these four Northeast Asian nations alone over the 1990–95 period, and an absolute scale of spending more than double that which the United States devoted to its huge Pacific forces. It was roughly as much as Russia spent on its armed forces worldwide.[16]

Asia's ongoing defense buildup, of course, is by no means limited to these four nations of Northeast Asia. Russia and Vietnam, not to mention North Korea, already have huge armed forces in being from previous buildups in the Pacific, especially during the 1970s and 1980s. Pyongyang, for example, added more than one thousand tanks to its arsenal during the 1980s, together with several thousand heavy artillery pieces. Russia meanwhile undertook a massive expansion of its Pacific fleet, giving it seventy-five major combat ships, including aircraft carriers, by 1990.[17] Vietnam accumulated huge stockpiles of military equipment at the end of the Vietnam War, to complement its already powerful army. Most of these huge Cold War–era deployments remain intact, although their combat readiness is clearly declining with time.

Table 7-2

ASIA ARMS AS THE WORLD CUTS BACK: POST-COLD WAR CHANGES IN DEFENSE BUDGETS (1990-95)

	1990	1991	1992	1993	1994	1995	% change 1990–95
U.S.	291.4	3.0	270.9	258.9	251.4	252.6	−13.3
USSR	116.7	—	—	—	—	—	—
Russia	—	—	85.9	75.1	78.5	62.8	−26.9[a]
Japan	28.7	32.7	35.9	39.7	42.1	53.8	+87.5
China (PRC)	6.1	6.1	6.7	7.3	6.7	7.5	+23.0[b]
Taiwan	8.7	9.3	10.3[c]	10.5	11.3	NA	+29.9[c]
South Korea	10.6	10.8	11.2	12.1	14.0	14.4	+35.8
North Korea[d]	5.3	2.4	2.1	2.2	2.3	2.2	−58.5
Vietnam[d]	NA	1.9	1.8	NA	0.9	0.9	NA
Indonesia	1.5	1.6	1.8	2.0	2.3	2.6	+73.3
Australia	7.0	7.1	7.0	7.0	6.9	7.4	−5.7

Sources: IISS, *The Military Balance,* 1990–96 eds.

Notes: Figures in $ billions unless otherwise indicated. Figures are for defense budgets, which indicate intended level of effort and for which more recent data are available, rather than defense expenditures. Data do not correct for exchange rate fluctuations and hence exaggerate some changes.
[a] 1992–94.
[b] IISS estimated actual Chinese expenditures are much higher: $18.8 billion for 1991, $24.3 billion for 1992, $27.4 billion for 1993, and $28.5 billion for 1994.
[c] 1990–94.
[d] Estimates.

Indonesia is expanding rapidly in percentage terms, with more than a 73 percent increase in defense spending over the 1990–95 period as it began deploying the entire East German navy.[18] Even the Philippines, following the sudden Chinese occupation in early 1995 of the aptly named Mischief Reef in the South China Sea, claimed by Manila, has launched a five-year, $2 billion buildup program, sharply expanding aircraft and patrol-boat purchases.[19] Although the heart of East Asian military expansion may be the northeast part of the region, it is a far broader and more complex, interactive dynamic across the area as a whole. Therein lies much of its danger for the future.

The Specter of a Naval Arms Race

Given East Asia's geography, as a region of vast ocean expanses, and its political economy, as an area intimately linked with the rest of the globe by energy sourcing and other forms of trade, it is not surprising that the sea is a major forum for Asia's emerging arms race competition. The most substantial and modern naval force in the region, apart from that of the United States, is clearly Japan's. Japan has more than one hundred maritime aircraft, including the largest fleet of PC-3 submarine surveillance aircraft in the world apart from the U.S. fleet, together with over sixty principal surface combatants and eighteen submarines.[20] Despite constitutional constraints, Japan is already, in accordance with a 1981 declaration encouraged by the Reagan administration, involved in maritime operations out to one thousand nautical miles from its shores.

Because the Japanese archipelago and dependencies like Okinawa and Iwo Jima are so spread out, Tokyo's naval defense perimeter sprawls across nearly a quarter of the entire North Pacific, reaching almost as far south as the Philippines. Within its geographic purview, significantly, are the environs of Taiwan, potentially among the most hotly contested waters in Asia over the coming decade or two, although the prospective lines of confrontation remain, from a Japanese perspective, distressingly uncertain. To better defend its huge expanse of maritime concerns under all possible contingencies as geostrategic realities in Asia change, Japan recently launched the first of at least three $1 billion destroyers equipped with U.S. Aegis radar surveillance and tracking systems. It is also planning to acquire tanker aircraft to extend the range of its maritime air coverage and has been considering for several years the acquisition of "defensive" aircraft carriers.[21]

A major question for the future is obviously the prospect for Japanese blue-water force projection, of which aircraft carriers would be a central element. Clearly, as suggested above, Japan's geostrategic and political-economic incentives for such capacity may very well grow stronger over the coming decade, particularly if China, as a major oil importer, develops major blue-water naval capacities, and if the U.S. presence in Asian and Indian Ocean waters significantly recedes. Acquisition of a full-scale blue-water navy would be enormously expensive for Japan (perhaps on the order of $30–40 billion U.S.) and time-consuming to build; it would need, for tactical reasons, to involve the acquisition of full aircraft-carrier battle groups and significant naval air power.

These reinforcing elements would be necessary to assure that the blue-water navy could protect itself, in a self-sufficient way, amid potentially hostile waters far from its native shores. Developing blue-water capacity would almost inevitably force Japan to revise its "no war" constitution, a politically delicate proposition. But Japanese shipbuilding and defense-electronics capabilities are both already among the strongest in the world. With 15 percent of global GNP, there is clearly no question of Japan's economic ability to gradually build its own blue-water navy when and if circumstances require.

Over the past decade, Japan has subtly, under encouragement initially from the United States, begun to develop substantial naval capacities. These could ultimately, especially under conditions of a broader Asian arms race, culminate in blue-water naval capacity slightly beyond the year 2000. Although Japan has so far consistently refrained from developing cruisers, aircraft carriers, and nuclear submarines, leaving destroyers the largest vessels in its navy, it has steadily expanded and redefined both the scale and the functions of its destroyers. They now displace nearly ten thousand tons under full-load conditions, which would have classified them as light cruisers even a decade ago.[22]

The Maritime Self-Defense Force is likewise moving incrementally toward amphibious capacities that at least preserve the option of aircraft carriers in the future. In particular, the MSDF's planned acquisition of several advanced-design amphibious landing ships may signal the beginning of such a force with advanced power-projection capabilities. The new ships will be 588 feet long, displace 8,900 tons, and have a maximum speed of twenty-two knots. More important, the upper deck will be capable of carrying amphibious vehicles and serving as a flight deck for helicopters or even vertical-takeoff-and-landing fighters, like the U.S. Marine Corps's Harrier jets.[23] The new ships will be a far cry from modern aircraft carriers like the Yokosuka-based U.S.S. *Independence*. But they will have, in a limited way, some of the same ambitious functions.

The China Factor

The major catalyst for deepening naval rivalry in East Asia clearly is China. With an economy that has averaged 9 percent real annual economic growth continuously since 1979, giving it now the third-largest GNP in the world, China manifestly commands the resources for major military-industrial expansion. With rising energy imports, a military complex exposed by the Gulf War as obsolete, and annoying affronts to great-power pride on its periphery, China clearly has the incentives to do so.

The Chinese navy still remains relatively small, with limited operational range and many old ships. Limited antiair defense, logistics, and antisubmarine warfare

(ASW) capacities are also still a distinct handicap.[24] Some of the People's Navy's initial steps toward blue-water power projection, such as its consideration and then rejection of the opportunity to buy the Ukrainian aircraft carrier *Vartag*, have been halting.

Yet China's navy does already possess some elements obviously designed for an offensive mission. It has a large amphibious force, which could be used against Taiwan as well as rival South China Sea island claimants, together with many destroyers and frigates boasting impressive missile armaments. China's naval air forces have also reportedly acquired air-to-air in-flight refueling capacity that could help enforce its sprawling territorial claims across the East and South China seas.

China, contrary to general conception, has a strong naval tradition. Indeed, the Middle Kingdom has oscillated over the centuries from a continental to a maritime strategic focus, as the locus of challenge to its national interests has likewise fluctuated over time. During the Han, Song, and Ming dynasties, in particular, China had a strong seaward orientation; at its height in the early fifteenth century, the Ming navy, for example, consisted of 3,500 vessels and ranged with confidence as far as the Red Sea and the east coast of Africa.[25] Even though periodic land-based threats, such as the Manchus in the seventeenth century and the Soviet Union and the Japanese in the twentieth, have oriented China back to continental preoccupations, a periodic cyclical oscillation back to the sea has continued. Just as Admiral Zheng He sailed the Indian Ocean in fleets of treasure ships four times the size of Columbus's *Santa Maria* more than five hundred years ago, it is highly possible that the shadow of China may loom large once again over sea routes between Asia and the Middle East in decades to come.

Among the clearest signs of rising Chinese interest in naval power projection far from China's shores are its military assistance to strategic Indian Ocean states, extending as far as Tanzania, and its accelerated program of air and naval base construction. China has reportedly acquired rights potentially leading to a naval presence on the Gulf of Martaban in Myanmar, only a few hundred miles northwest of the strategic Strait of Malacca, through which virtually all of its own and most of all East Asia's rising oil imports in coming years must flow.

Following South China Sea clashes with Vietnam in 1988, China has constructed a major air base at Woody Island in the Paracels, and it is building another on its southern island of Hainan, from which its new long-range Su-27 Flanker fighters will operate. From the PLA's third major air base, at Zhanjiang in southern China, which is also headquarters of China's South Sea fleet, newly deployed tanker aircraft will enable Chinese combat planes to operate for extended distances over strategic sea-lanes and offshore oil fields in the South China Sea. There they can enforce disputed territorial claims that stretch a thousand miles south from China's shores, nearly to the coasts of Malaysia and Indonesia. Acquisition of the beautiful protected harbor at Hong Kong in 1997 should strengthen China still further.

China has also begun the long process of acquiring the core of a blue-water navy. This could have important implications both for the South China Sea conflict and relations with Japan, especially toward the year 2000 and beyond. China has been training aircraft carrier pilots and working on carrier technology, as well as acquiring both prospective carrier planes and attack submarines necessary for protecting a carrier task force. Within a decade it should have multiple operational aircraft carriers.[26]

As China expands its naval sway, island Taiwan, only ninety miles across the Taiwan Strait, is directly challenged. Many of Taiwan's greatest fears lie on the sea, in the form of a possible mainland amphibious attack or blockade. Maintaining an upper hand in naval matters is therefore crucial. Taiwan is purchasing sixteen Lafayette-class frigates from France for $4.8 billion, and building eight U.S. Perry-class frigates in its own shipyards, which will carry home-built Hsiung Feng II surface-to-surface missiles. A German consortium is providing Taipei with ten more frigates for $12 billion. Including huge pending orders for submarines, Taiwan appears likely to spend $40 billion on armaments over the coming decade.[27]

As Northeast Asia's political-economic dependence on sea routes to the Persian Gulf rises over the coming decade, both the leverage and the vulnerability of Southeast Asian states commanding the narrow, strategic straits through which such routes pass are also rising. Not surprisingly, the "porpoises" of the region, such as Indonesia, Malaysia, Thailand, and even Singapore, are also getting swept up in the broader regional arms race.[28] All of them are transforming their naval capabilities from what have been essentially coastal forces into navies with greater range and a significantly broader variety of capabilities. Indonesia, for example, purchased the entire East German navy, shortly after the collapse of Communism there, and inaugurated an active buildup program that will deploy as many as twenty FG-90-class frigates by the late 1990s. Indonesia, Malaysia, Singapore, and Thailand are all investing in mobile combat forces and long-range bomber/attack squadrons, while also acquiring Harpoon and/or Exocet antiship missiles. Thailand has also commissioned a 9,500-ton helicopter carrier, built in Spain.[29]

Both Southeast and Northeast Asian nations, finally, are expanding their submarine fleets, responding to the accelerating naval buildup and the increasing strategic significance of regional sea routes. Only the United States, Russia, and China have nuclear submarines. Yet virtually all the other nations of the region are busy expanding or upgrading their conventional undersea fleets.

Japan has a venerable submarine tradition; indeed, its navy was the first in the world to use submarines in combat, during the Russo-Japanese War of 1904–5.[30] During World War II it produced some of the most technically advanced subs of the war, although it failed to use them to good advantage. Japan during 1990–92 introduced its modern Harushio-class diesel-electric sub, displacing 2,450 tons, highly automated, armed with missiles and torpedoes, and technically well received.

By 1995, Japan had eighteen submarines in all.[31] Taiwan had contracts with a German consortium for ten submarines. Indonesia, Malaysia, Singapore, and Thailand all had likewise moved to upgrade their submarine fleets, in a major new intensification of arms rivalries in Asia.

Arms Race Asia Takes to the Air

Closely related to the emerging naval arms competition in East Asia, and paralleling it in intensity, has been an escalating buildup of air force and missile capabilities. As in the naval area, redefinition of Chinese strategic requirements, as China's economy grows and its political relationships with neighbors change, drives much of the broader weapons competition within the region. Three developments have simultaneously traumatized and stimulated China: the Gulf War of 1991, the collapse of the Soviet Union late that same year, and Taiwan's huge, ongoing $40 billion weapons procurement program.

The Gulf War traumatized China, as it so starkly revealed the bankruptcy of the Soviet weapons technology that the Iraqis and the Chinese themselves possessed, as well as the devastating power of American defense electronics and communications. Deprived of leverage on the global scene by collapse of the Soviet Union and the resulting devaluation of the "China card," China found itself geopolitically naked, and it began to see immediate military modernization as a national-security imperative. There were economic pressures as well: China's lucrative profits from arms sales plummeted 80 percent between 1988 and 1992, as its Third World clients perceived the same vulnerability of Chinese weaponry, as compared to American arms, that was sensed in Beijing following the Gulf War.[32]

As a vulnerable China cast around for reassurance and support of its military modernization, a collapsing Soviet Union was just embarking on the greatest military yard sale in history. Not surprisingly, the Chinese, flush with hard currency drawn from their soaring, multibillion-dollar transpacific trade surpluses, stocked up. They began by buying dozens of Su-27 fighters and made plans to produce their own version of Russia's top-of-the-line MiG-31 strategic interceptor, using a small army of fifteen hundred Russian engineers and technicians. Hundreds more were put on retainer, creating an elaborate E-mail network between Russian and Chinese defense research institutes that has since accelerated the development of Chinese aerospace and nuclear programs.

China's air force modernization program was by no means defensive. Apart from the MiG-31s to be coproduced, China also reportedly acquired air-to-air refueling technology from Iran, which had gotten it, in turn, from the United States during the reign of the shah. It also purchased Tu-22 long-range bombers, IL-76 military transports, S-300 ground-based antiballistic missiles, and A-50 airborne warning and control planes from the Russians. Not surprisingly, this buildup, coupled with the naval expansion outlined above, deeply alarmed the Taiwanese, accelerating still further their ambitious armament program. Taipei gazed across the ninety-mile-wide Taiwan Strait at a sworn enemy with a three-million-man army, nuclear weapons, and a fixation about "liberating" their island. This long-standing adversary was also buying billions of dollars' worth of new weapons from Russia and setting up Sino-Russian military cooperation programs that could funnel much of Moscow's superpower nuclear and aerospace know-how to Beijing in coming years. Not surprisingly, Taipei grew ever more alarmed. With nearly $90 billion in foreign exchange, the largest government reserves in the world, money was not an issue.

To counter the prospect of Chinese MiG-31s, Taiwan considered fighters from throughout the world, including the U.S. F-16, the Swedish Grippen, the French Mirage 2000, the Russian MiG-29, and the Israeli Kfir. Amid a desperate reelection campaign, George Bush decided in the summer of 1992 to sell Taiwan 150 F-16s for $6 billion, announcing the deal on September 2 at the F-16 plant in Fort Worth, where 5,800 of 20,000 jobs were threatened with layoffs because of shrinking orders. The deal contravened the Reagan administration's decade-old agreement with China to freeze U.S. weapons exports to Taiwan at $700 million a year, but it helped Bush carry Texas, even if not the presidential election, in November. On the heels of the F-16 deal, Taiwan also signed a $3.8 billion contract with Dassault to deliver sixty French Mirage 2000-5 multipurpose advanced combat jets to Taiwan as well. Both Taiwan and China will thus be steadily adding top-of-the-line U.S., French, and Russian fighters to their expanding respective arsenals well into the next century.

The arms race in the air is also spiraling into Southeast Asia. Singapore, Indonesia, and Thailand are all assembling, in response to one another and to the uncertain situation in the South China Sea, major fleets of F-16s. Malaysia in June 1994 purchased eighteen MiG-29s, marking Russia's entry as supplier to non-Communist Southeast Asian countries. Military-run Myanmar has diversified its acquisitions, purchasing both Chinese F-6 and F-7 fighters and the Yugoslavian Super Galeb aircraft. Hanoi, of course, already has one of the largest air forces in the world, much of it U.S. equipment captured with the fall of Saigon in 1975, and still largely housed in U.S.-built, mortar-proof concrete hangars at the sprawling former Pentagon East, now Tan Son Nhut Airport in Ho Chi Minh City.

Technological change is making advanced missiles an increasingly central and indispensable element in Asian military arsenals. As noted above, all the ASEAN air forces and navies are now being equipped with large numbers of either French Exocet antiship missiles, so destructive in the Falklands conflict, the American Harpoon, or the British Hawk. China has its own extensive missile production program and has been exporting Silkworms and "M-class" surface-to-surface missiles aggressively to the Middle East. So has North Korea, whose Rodong 2 missile being jointly developed with Iran will have an eight-hundred-mile range, making it capable of hitting Israel from Iranian territory.[33]

Both Taiwan and South Korea likewise have their own advanced, domestically produced missiles, such as Taiwan's Skybow SAM and Cheng Kung naval missile. Taiwan also reportedly deploys land-based Hsiung Feng II missiles on small islands in the Taiwan Strait, targeted on Fujian province, where they confront China's M-9s and M-11s.[34] These parallel deployments assure that any future conflict in that strategic waterway will be a hair-trigger contest pitting sophisticated, locally produced missiles against one another on both sides.

Most of the spiraling arms competition in Asia is waged, as suggested above, either in the air or on the sea. There are few of the great, tense land boundaries, like the notorious, long-standing open lines of confrontation between NATO and the Warsaw Pact at the Fulda Gap in Cold War Germany. Korea, of course, is the one great exception.

Korea as Arena to Conflict

Stimulated by big-power rivalries, a bitter, persistent arms race has been continuing in Korea—on land and sea and in the air—ever since North and South were formally partitioned in 1948. This exploded once before, of course, into open armed conflict, when the North driver, would and launched the Korean War in 1950. Despite the armistice of 1953, tensions continued across the four decades and more that followed, producing today's massive deployments and the specter of nuclear confrontation on the tortured Korean peninsula.

One important recent development has been the substantial buildup of South Korean conventional naval and air power, on the one hand, and North Korea's attempts to balance this. The North has combined its nuclear program with a determined effort to develop advanced missiles, and with highly adventurous forward deployments along the DMZ. In the background is the shifting balance of economic power, at once potent and destabilizing, across the 38th parallel. Despite heavy new acquisitions over the past decade and defense spending that rose more than 63 percent between 1985 and 1992, the South Korean defense budget actually fell from 5.1 to 3.8 percent of GNP during that period, thanks to buoyant economic growth. North Korea, conversely, faced with economic stagnation and a GNP that declined every year throughout the early 1990s, was only able to increase its military budget one-third as fast, despite a deepening military commitment that took defense spending to 25.7 percent of GNP—virtually the highest in the world relative to economic output—by 1992. Thus Pyongyang's movement toward nuclear weapons and reckless forward deployments, in a desperate attempt to retain eroding military leverage with its adversaries, is hardly surprising.

In the face of North Korean militancy, as well as uncertainties elsewhere in Asia, South Korea has steadily expanded its own military capabilities, to the point that they now constitute a rather formidable force in international terms. Increasingly, the emphasis has been shifting from land to sea defense, as Seoul looks beyond the North-South confrontation to longer-term issues of the regional power balance and sea-lane security, many of which were spelled out in its defense ministry's controversial 1992–93 white paper. Geostrategically an island, cut off from allies by oceans and the forbidding DMZ, South Korea imports virtually all of its crude oil from the Middle East, in effective competition with Japan and China. So there is arguable strategic logic to Seoul's new emphasis on sea power.

The South is in the midst of a major long-term naval expansion program that may ultimately bring it some significant blue-water naval capacity. Over the next decade it plans to acquire up to seventeen new destroyers, twenty frigates, fifty corvettes, and up to sixty-eight fast patrol boats, with some possibility as well of a small aircraft carrier. Indeed, in mid-1995 a South Korean trading firm purchased two Kiev-class aircraft carriers, the *Minsk* and the *Novorossiysk*, from the Russian Pacific fleet at Vladivostok, while stoutly claiming that it was buying them only for scrap.[35] With the two 40,000-ton carriers holding a total scrap value in the vicinity of $10 million, many suspected broader geostrategic intentions at work, if only an interest in the technology.

In the air, South Korea is acquiring 120 new F-16s, to add to the thirty-six that it already deploys. It also is buying British Hawks, as well as American Patriot antimissile missiles, similar to those deployed in Saudi Arabia and Israel during the Gulf War. Its buildup thus is a most comprehensive one.

Japan and the Balance of Power

Surrounding Japan, as suggested in Chapter 2, is thus an arc of insecurities and regional rivalries: to its north the rusting hulk of the Russian Pacific fleet, still deadly with nearly twenty operational nuclear ballistic missile submarines; to its northwest the North Korean Rodong 1 missile and the Korean peninsular arms race; to the west a modernizing Chinese People's Liberation Army looking increasingly to sea; and to the southwest an escalating Taiwan Strait military confrontation. Farther beyond the sea horizon, or the *suiheisen*, as Japanese call it, lie the deceptively placid South China Sea and the narrow, strategic straits of Malacca and Lombok. Through those choke points flow 70 percent of Japan's economic lifeblood, petroleum, together with an increasing volume of Korean and Chinese oil. Fueled by economic growth, energy uncertainties, and political confrontation, the shadows of rising arms competition flicker on Asian shores. In a world of rising economic interdependence, those shores are no longer so distant from Japan's own.

Ultimately Asia's dangerous new power game, with the specter of a heavily armed and unified China and Korea on its doorstep that it presents in worst-case scenarios, threatens to destabilize Japan's traditional low-posture military orientation. It also threatens to provoke, over the long run, a serious arms race, centering on Japan and China, that could have global implications. Given Asia's affluence, technological sophistication, and rapid economic growth, its local military rivalries could provoke a buildup among Asia's strongest powers ultimately formidable enough to threaten even the United States, and thus pose a direct, albeit very long term, challenge to Pacific Defense.

Across the past half century, Pax Americana and the Cold War bipolar confrontation have inhibited and defused Asian balance-of-power rivalries that reared their ugly heads so frequently and disastrously before 1945. Clearly the Cold War cannot play the inhibiting role in future that it did over the past fifty years. The American role in the region, including the U.S. security presence on Asian soil and in Asian waters, looms large as a guarantor that the tensions implicit in the emerging balance-of-power game within the region will not escalate. Yet that role is in turn bound up with the American—and indeed the Asian—stake in a global system of interlocking economic and security interests. In a post–Cold War, post-GATT world where regionalism holds increasing sway in trade and investment, what common stakes are being defended? Who pays the costs, and who sets the details of security deployments? What common incentives and sentiments do America and Asia have to support each other, and against whom? It is to such delicate, troublesome, and vital topics—fundamental to Pacific Defense under almost any calculus—that we now must turn.

ASIA AND THE TWILIGHT OF GLOBALISM

Pacific Defense is not a narrow matter of military deployments alone. Security in the broadest and most important sense, in an interdependent world where no one nation stands omnipotent, is a matter of fostering alliance ties to others based on common interests. For America this binding process has traditionally had an economic dimension, based on the ideal of global free markets and unimpeded flows of capital.

Sobered by the turbulence of global depression and war across the three previous decades, financial statesmen from all over the Allied nations gathered at Bretton Woods, New Hampshire, in the late summer of 1944. It was just after D-Day, in the wary euphoria of impending victory over Hitler, that they met to plan for the long-term future. Their modest objective: to create a unified global economy, spurring trade and investment, and thus ensuring global peace. Their vehicles: the International Monetary Fund, the World Bank, the General Agreement on Trade and Tariffs (GATT), and ultimately the United Nations itself.

Globalism, as John Maynard Keynes and Harry Dexter White conceived it, succeeded for years beyond their wildest dreams. Within two years of Bretton Woods, following closely upon the surrender of Germany and Japan, the IMF and the World Bank were in being. By the end of the 1950s, all the major currencies of Western Europe and North America—accounting for over 80 percent of global trade—had become convertible. Tariffs and nontariff barriers were on their way steadily down. By 1964, Japan, Asia's rising economic powerhouse, had joined the Organization for Economic Cooperation and Development (OECD), and begun to subscribe to the codes of the industrialized world. By 1986 the world had completed its second major round of global liberalization talks and begun work at Punta del Este, Uruguay, on a third one.

For most of Asia, especially the collection of smaller non-Communist nations (Japan, South Korea, Taiwan, and ASEAN, in particular) perched precariously around the periphery of China, globalism was an economic godsend that bound them firmly to American conceptions of Pacific Defense. Globalism opened first American and then broader international markets to Asian exports and loosened a flow of investment from multinational firms in relentless search of cost-effective production bases. These Asia provided in profusion. Both Asian growth and Asian trade surpluses, in the aggregate, began to accelerate sharply, especially from the 1960s on. During the past decade, China has also gradually begun to join, however imperfectly and asymmetrically, this globalist economic framework.

Trade and financial globalism have had their functional analogue in the security sphere: the web of Cold War mutual security treaties that protected the members against feared Soviet or Chinese aggression. To be sure, in security matters the globalist analogy was imperfect: NATO covered only the Atlantic, while Pacific security alliances were largely bilateral. Apart from the 1951 U.S.–Japan Mutual Security Treaty, signed at the San Francisco Opera House two hours after the peace with Japan, the ANZUS pact with Australia and New Zealand, a bilateral treaty with South Korea, and the SEATO alliance in Southeast Asia completed the formal, narrowly military architecture of Pacific Defense.

The overall arrangement was, as James Baker put it, a "hub and spokes" structure, with the spokes emanating from the hub of Washington.[1] Here we call it, after its origins in the U.S.–Japan alliance at its core, the San Francisco system. Felicitously, all the key non-Communist economic and security arrangements around the world had the United States as a central partner: they transcended both the Atlantic and Pacific, linking the U.S. securely to allies on either shore. Indeed, they forged a reinforcing structure of economic and military power whose unobtrusive dynamism finally became clear with the collapse of the Berlin Wall and literally all of the West's European Communist adversaries in twenty-six short, dramatic months during 1989–91.

American-led globalism in the security sphere, like its economic counterpart, likewise had fateful consequences for Asia. To be sure, it at times constrained the foreign policy options of American Pacific allies, effectively isolating them from the mainland Chinese heartland of their region, to which many of them had been intimately linked before World War II. Yet globalism in its security dimension also had the outstanding virtue of muting the mutual fears and insecurities within the region that had been so pronounced before 1945, and that were badly exacerbated by the ravages of World War II. The value of this globalism, in the form of the U.S. security umbrella and troop presence in Asia, was attested to not only by the Asian business world and the U.S. State Department, but ultimately by no less a traditional foe of U.S. hegemony than Chinese prime minister Chou En-lai. The durable geopolitical logic for many Asians of a U.S.–led globalism in security matters had grown so strong by the mid-1990s that even the Vietnamese were implicitly endorsing it, inviting U.S. Pacific Fleet commander Richard Macke to Hanoi and informally offering base facilities at Cam Ranh Bay, which they had fought bitterly to wrest from the United States only two decades before.

The transoceanic security frameworks of Cold War years continue to persist, at least in form. Yet the economic and even cultural foundations upon which they rest are in the throes of arguably profound transformation. In the twilight of the Communist and capitalist universalism that the Cold War so encouraged, more limited conceptions of identity and interest now rear their heads. The "one world" of Keynes, Roosevelt, Churchill, Lenin, and Mao has dissolved into a much more parochial set of equations, with implications that could in time trouble the security frameworks that originally gave birth to economic globalism in the first place.

The twilight of globalism has particularly fateful implications for East Asia. This trend strikes at its prosperity, by threatening markets for the highly competitive products of Asian industry that have developed throughout the world. It also threatens its stability, already beset by domestic tensions, by eroding the incentives of America to remaining militarily committed in Asia, and thereby inhibiting the balance-of-power rivalries that could otherwise seriously divide this region, as just suggested in Chapter 7.

The Decline of Economic Globalism

Decay in the globalist structure first materialized in the sphere of finance. Rising current account surpluses in resurgent Germany and Japan, coupled with massive American deficits incurred to simultaneously wage the Vietnam War and the domestic War on Poverty, destroyed the system of fixed exchange rates and dollar convertibility into gold during the early 1970s. These twin pillars were the heart of the Bretton Woods system.

Markets have, to be sure, been able to sustain—indeed, to deepen—global financial interdependence in ensuing years. A de facto floating-rate system has remained in place since 1973, withstanding and helping to ease such tremendous political-economic strains as the two oil shocks of the 1970s, the Iraqi invasion of Kuwait, and the collapse of the Soviet Union. Today interdependent global currency markets in New York, London, and Tokyo handle well over $1 trillion a day in transactions with remarkable ease.

Yet despite sporadic efforts at cooperation, like the 1978 Bonn G-7 summit and the 1985 Plaza exchange-rate accord, the major nations have never been able to restore the unity and orderliness of Bretton Woods. At times their policy differences have triggered global instability, as when U.S.–West German feuding provoked a world financial crash in October 1987. The debt crisis of the early 1980s, like its relative the peso crisis of 1995, also threatened globalism, by inhibiting investment in the developing world and thus deepening North-South income disparities.

East Asia, especially Japan, has suffered from the eclipse of an ordered global exchange-rate system and the ensuing volatility in global markets. These developments triggered a more than fourfold appreciation of the yen between 1971 and 1995, together with substantial appreciations of the South Korean won and the Taiwanese NT dollar as well. To be sure, consumers in countries with such appreciating currencies secured some welfare gains. Yet these were often absorbed by complex, inefficient distribution systems, collusive local marketing practices, and other nontariff trade barriers. Asian investors, meanwhile, suffered huge portfolio losses on their global investments. After a surge into American and other global markets in the late 1980s, by the mid-1990s they were banking their funds much closer to home. America and Europe clearly held less attraction than prosperous Asia itself.

Globalism in trade has confronted even greater pressures. A few countries, concentrated largely in East Asia, have profited enormously from the global system sustaining huge and rising trade surpluses. Japan and China alone, for example, recorded a combined surplus with the United States of nearly $100 billion in 1994. Yet the majority of nations have done badly, making them both less willing to support free trade and more conscious of the protectionist sins of others. The Middle East, Latin America, and much of Africa, in particular, have been plagued by political instability, compounding the difficulties of pursuing open, consistent trade policies. Foreign investors and local entrepreneurs have both been further discouraged by misaligned exchange rates, corruption, unfavorable regulatory regimes, and pervasive, inefficient state socialism.

Except for a few newly industrializing economies (NIEs), generally adjacent to the three core advanced regions (North America, Western Europe, and East

Asia), most of the so-called Third World has been a stagnant backwater for a decade and more, little touched by broader waves of international economic advance. The bulk of world trade has been conducted within or among these three global cores. And yet despite progressive attempts at formal liberalization on a global basis through the Kennedy, Tokyo, and Uruguay rounds, world trade grew progressively more regionalized across the 1980s and early 1990s within Europe, North America, and East Asia, as Table 8-1 points out.

The figures in Table 8-1, it should be noted, do not include rapidly growing China. It has become a magnet for trade in recent years, especially for other East Asians. Indeed, more than two-thirds of China's trade is with nations of the Pacific other than the United States, a figure which has been remarkably stable throughout the past decade.[2] When China, whose largest trading partners are Hong Kong and Japan, is added to the calculus of regionalist tendencies, trade interdependence within East Asia becomes significantly greater than that within North America—probably on the order of 45 percent of total trade by the mid-1990s.[3] Yet even this figure may not sufficiently account for the substantial but informal border trade and other politically ambiguous economic relationships across old Cold War boundaries (such as China-Taiwan, China-Vietnam, North–South Korea, and China-Russia). These have begun to expand rapidly in East Asia over the past five years.

Mutually reinforcing political and economic forces are fueling a steady regionalization of trade among the advanced industrial nations. In Europe the watershed political development was clearly the signing of the Treaty of Rome, creating the European Economic Community in 1957; the resulting customs union, with common tariff walls among France, West Germany, Italy, and the Benelux nations, was followed by a series of measures stimulating intraregional trade. These culminated on January 1, 1993, in a "single market," with unified investment, environment, and regulatory policies among the twelve nations of the expanded community. Europe has also created a European Parliament and European passports, while producing an integrated military force, the Euro Corps, as well as trans-European transport and communications infrastructure. The newly christened European Union is also expanding steadily to the north and east, with the admission of Austria, Sweden, and Finland in 1995 and the prospect of additions from the former Communist bloc well within a decade.

The United States, Canada, and Mexico responded to Europe's initiatives toward integration with the North American Free Trade Area (NAFTA), likewise a comprehensive customs union. This was formally inaugurated on January 1, 1994, just a year after the birth of the European Union. North America has not

Table 8-1

THE RISE OF INTRAREGIONAL TRADE

	East Asia[a]	North America[b]	European Community[c]
1980	33.8	31.5	52.4
1986	34.6	33.0	57.5
1990	38.4	37.3	63.4

Source: International Monetary Fund, *Direction of Trade Statistics.*
Note: Figures are percentages of total trade.
[a]Includes Japan, South Korea, Taiwan, Hong Kong, and ASEAN.
[b]U.S., Canada, and Mexico.
[c]Twelve members of the European Community.

yet attempted the ambitious transnational political institutions of Europe. But even in the face of setbacks like the peso crisis of early 1995, regional trade interdependence is clearly deepening. And steps like the 1994 Miami summit commitment to end inter-American tariffs early in the next century are accelerating that integration still further.

Asia, unlike Europe and to a lesser extent North America, has not pursued elaborate formal approaches to regional integration. To be sure, there are some active regional bodies: ASEAN, founded in 1967; the semiofficial Pacific Economic Cooperation Conference (PECC), founded in 1980; the governmental-level Asian-Pacific Economic Community (APEC) group, established in 1992; and the Asian Regional Forum (ARF), a security deliberation body created in 1994, to note some of the most important. Ambitious long-term trade liberalization goals, including free trade among advanced Pacific nations by the year 2010, have been set by the leaders of the APEC nations.

Yet few implementation mechanisms are in place. None of the Pacific regional organizations, including APEC, has significant actual regulatory power. Industrial East Asia, in contrast to other analogous advanced regions, has no customs union or systematic economic coordination mechanism. Yet as indicated above, it nevertheless has a vigorous and rapidly growing pattern of internal trade, which deflects the interest of its members increasingly from globalist concerns.

From a global political-economic perspective, the most fateful recent casualty of the decline of globalism in international trade is the waning importance of the American market to East Asians. Between 1986 and 1992 the share of East Asian exports flowing to the United States fell a full ten points, from 34 to 24 percent.[4] As Asian economic dependence on the United States declines, so often does U.S. influence, eroding a fundamental pillar of transpacific relations as we have known them since World War II.

Mutual Attractions Within Asia

One key reason for the recent intensification of regional trade within East Asia has been the rapid expansion of the two powerful "locomotives" of the western Pacific: first Japan, and then China. After completing its postwar recovery, Japan during the 1960s and early 1970s achieved steady, double-digit growth. Yet it took in only modest levels of imports, because of low domestic consumption. During 1977–78 and once again during the "bubble" expansionary period of the late 1980s, however, Japan was more successful at pulling in imports, cutting its huge current-account surplus and thus spurring the Pacific regional economy.

In deepening its Asian trade ties, Japan by no means forged a coprosperity sphere. But it did erect a cohesive manufacturing complex in Southeast Asia, producing consumer electronics, cameras, air conditioners, and simple communications equipment. This sharply deepened the trading links of key Asian nations to it, as moths to a flame. Between 1986 and 1992, Thailand, Indonesia, the Philippines, and Singapore all sharply expanded Japan's share of their total trade, while the American share waned.[5] For Japan itself, the shift to trade regionalism was even more pronounced. Trade with the United States fell sharply to 26 percent of Japan's total, while Asia's share soared by more than a third to 32 percent, eclipsing the United States for the first time in half a century.[6]

Intra-Asian trade to and from Japan has grown more lucrative, as well as more pervasive. Consumers in high-growth Asia are becoming more prosperous and can afford to pay more for quality Japanese goods. The values of most currencies in non-Communist Asia, including Taiwan, South Korea, Singapore, Malaysia, and even Thailand, have also appreciated against the U.S. dollar across the past decade, increasing the allure of these markets as compared to the United States. New post–Cold War production opportunities are also opening in China, Vietnam, and potentially even North Korea. Those countries, whose currencies remain weak, offer high-margin opportunities for Japanese firms to export back to Japan. Such opportunities grow ever more lucrative with the yen's appreciation and the prospect of exporting back into an affluent Japanese market.

China, since Deng Xiaoping's economic liberalization began in 1979, has been an even more successful and persistent, albeit smaller, locomotive for Asian regional trading partners than has Japan. Averaging 9 percent real GNP growth for the past fifteen years, China has not, in contrast to Japan, contracted huge trade surpluses that would be deflationary for trading partners in the region. To the contrary, Chinese trade, like Taiwanese or Korean, has been more consistently balanced with the world as a whole, even though China has begun to accumulate large bilateral surpluses with the United States.

As China's economy surged in the early 1990s, it became a potent locomotive for much of Asia. In 1993, for example, Chinese imports grew 29 percent, based on real economic growth of 13.4 percent, as trade surged to nearly $196 billion and a yawning trade deficit of over $12 billion materialized.[7] For some neighbors, the economic stimulus from China was very powerful: South Korean exports to China, for example, doubled in a single year during 1993, and China served as Korea's only buoyant international market. Japan, Singapore, and Taiwan, among others, have also found China an increasingly important trade and investment alternative to the United States, especially since 1992.

The rising importance of Asia's two new locomotives to the region's economic development—the very force that is weaning the region from the United States—has ironically been achieved at America's expense. China, in particular, is growing through heavy surpluses with the United States. These totaled nearly $35 billion in 1994, up from virtually zero in the mid-1980s. The Japanese surplus with America—at over $60 billion annually—looms even larger, although it is growing less rapidly. Together Japan and China account for more than three-quarters of the entire U.S. global trade deficit, an amount more than twice as large as all American oil imports.[8]

There are also smaller locomotives in Asia whose impact on intraregional trade, investment, and growth should not be ignored. South Korea and Taiwan are both nations of considerable economic scale, with a combined GNP larger than Canada's. And both are growing at 7–8 percent annually. Both nations' currencies are steadily, if slowly, appreciating against the dollar, and their share of global GNP is rising. Looking to the future, Korea, in particular, seems likely to rise in importance as a regional locomotive as it moves toward reunification and North Korea enters the Asian regional economy. Vietnam, with its rich natural resources and tenacious population of more than 70 million, also has potential, especially in the context of an increasingly invigorated ASEAN.

Regionalism and East Asian Trading Networks

Since East Asia is the most rapidly growing region on earth, with awesome economic prospects, nations from throughout the world desire to ship there. Yet despite Asia's massive collective trade surplus with the broader world, totaling well over $100 billion in 1994, its people buy increasingly from other Asians rather than from North America or Europe, with deflationary consequences for the West.

To understand these huge and persistent surpluses, one must look not only at Asian competitiveness, but also at the human dimension: Asian trading networks.

Business in East Asia is immensely personalistic: merchants trade with those whom they know. And the people they know and deal with above all are the local trading specialists of the region: the *sōgō shōsha* of Japan, such as Mitsubishi, Itoh Chu, Sumitomo, and Nisshō Iwai; the *chaebol* of Korea, like Samsung, Hyundai, and Daewoo; and the smaller, often family-based Chinese merchant firms.

The Japanese and Korean trading companies, in particular, are institutionally tied to broader conglomerates with a manufacturing arm. Although pragmatic and responsive to long-term market logic, these traders not surprisingly often favor firms and products from within the Asian region. And they retain enormous roles in foreign trade. Japan's nine largest *sōgō shōsha*, for example, are said to jointly handle 50 percent of Japan's entire export and import volume, although their role is limited in consumer sectors like autos and electronics that have generated the bulk of Japan's massive recent trade surpluses. In 1993, total sales of the nine *sōgō shōsha* came to nearly $1 trillion. Of this huge sum nearly a quarter was offshore trade, much of it in Asia, that did not touch Japanese shores.[9]

The major exception to the predominance of East Asian networks in Asia's trade with the world is the activity of large Western, particularly American, distributors, like Sears, Kmart, and Bloomingdale's. Taking advantage of low labor costs and relatively high product quality, they have sourced heavily in Asia for a quarter century and are now especially active in southern China. Although they are important in consumer products, their role is rather limited in other areas, however. In high-growth capital equipment and plant-export markets, for example, the Western role in product sourcing and distribution in Pacific markets is far less. Despite a few exceptions like General Electric, which boasts a corporate trade surplus of well over $1 billion with even Japan, all too many Western companies in the heavy industrial area are leaving the field to Asian competitors.

Investment Flows Closer to Home

The shadow of regionalism is deepening across the globe with respect to investment as well as trade. This can be seen most clearly, and most importantly from the perspective of this volume, by contrasting patterns of U.S. and Japanese investment since the early 1980s. There was, for example, an enormous existing stock of almost $240 billion in American investment in Western Europe at the end of 1992.[10] This was the heritage of a transatlantic era in the 1960s and 1970s, when a surge of American capital into the emerging European Economic Community dwarfed U.S. investment in Asia and even in Latin America. Yet by the late 1980s and the early 1990s that wave had crested. American investment was growing much more quickly in Mexico and Canada—the new U.S. partners within NAFTA—than in either Asia or Europe.

A parallel—indeed, even stronger—regionalist tendency is clear in Japanese direct foreign investment over the same period. Japan has an overwhelmingly large stake (almost $180 billion, even on a historical-cost basis) in the United States—the heritage of heavy commitments to the massive, expanding American market of the strong-dollar 1980s.[11] But the share of new Japanese foreign investment going to the United States has fallen by half since 1989, to less than $18 billion in 1994.[12]

Many of the largest and most visible Japanese investments in the United States have been plagued by disastrously poor profitability that has discouraged other prospective entrants. Sony, for example, which bought Columbia Pictures for $3.4 billion in 1989, recorded a huge operating loss of over $500 million in 1994 on its U.S. entertainment operations, which almost destabilized the parent company.[13] For Matsushita Electric, which paid $6.6 billion for MCA in 1990, even the windfall profits of *Jurassic Park* were not enough to compensate for the headaches of dealing with its newly "acquired" Hollywood management and the uncertain allegiance of Steven Spielberg.[14]

American real estate investments have also cost the Japanese dearly, prompting another steady retreat back across the Pacific. Overall, Japanese companies sold off $5 billion in U.S. real estate during 1994 and almost $10 billion in 1995, much of it at steep losses, especially when calculated in terms of yen.[15] Mitsubishi Estate, for example, by the end of 1994 had lost $500 million on its highly publicized 80 percent stake in Rockefeller Center. The pleasure of housing NBC Television, Time-Life, the Associated Press, Radio City Music Hall, and the Rainbow Room under its roof was by any measure faint compensation, triggering Mitsubishi's May 1995 decision to initiate bankruptcy proceedings.[16]

Asia is growing increasingly attractive to Japanese investors, with its low labor costs, diligent workers, familiar cultures, and absence of abortionist lawyers and lobbyists. Between 1989 and 1994, Asia's share of new Japanese investment overseas rose by more than half, from 12 to 18.5 percent of Japan's total.[17] Japan's cumulative investment in Asia now totals well over $60 billion, or more than triple that of 1985. In reality, Japan's actual stake in Asia may even be substantially larger, as much such investment is funded through internal resources or locally and hence fails to show up in the statistics.

Several other Asian nations, apart from Japan, have also become major capital exporters since the mid-1980s, including South Korea, Taiwan, and Singapore. Together with Japan, they held $253 billion in gold and foreign exchange reserves by the end of 1993,[18] or well over half the global total. Most important of these diminutive financial giants has been Taiwan, whose foreign exchange reserves at the end of 1994 were, at $92 billion, next only to those of the global leader, Japan. Although the local Bank of China holds most foreign exchange proceeds, as insurance against its delicate political situation, Taiwan's entrepreneurs have also invested heavily abroad.

Although during the mid-1980s much Asian investment went into real estate and resource plays in the United States, in more recent years it too has flowed heavily back to Asia itself. Since 1987, more than four thousand Taiwanese companies have set up operations in Southeast Asia alone, pouring over $25 billion, or nearly half their total global investment, into the region.[19] One typical commitment, symbolic of changing times, has been $20 million to build an industrial park, a tourist resort, and a Federal Express transshipment center on the site of the former U.S. naval base at Subic Bay in the Philippines.[20]

Encouraging investment within Asia (and inhibiting diversion to other parts of the world) is the presence of some twenty million overseas Chinese in the region, concentrated in Southeast Asia. Making secretive deals with far-flung relatives and friends, much as Jewish and Indian traders in the Middle East have traditionally done, these nimble Chinese expatriates neutralize risk and operate effectively in uncertain economic environments from Vietnam and Cambodia to the Russian Far East. Taiwanese companies also poured over $10 billion into fourteen thousand projects in mainland China during 1987–93, for similar reasons.[21] Well over half of this capital was supplied by small firms, often family businesses.[22]

American investment in Asia, to be sure, is rising moderately, attracted by the highest economic growth rates, and some of the highest profitability, in the world. Between 1989 and 1992, for example, U.S. investment in the region, including Japan, rose 40 percent to $78.2 billion, on a historical cost basis.[23] Income also rose 26.5 percent to $11.4 billion, with profit margins nearly 40 percent higher than the global average for American firms. Clearly the United States has an economic stake in Asia, and it is an attractive one.

Yet that American stake—and to an even greater degree those of erstwhile European colonial masters—in Asia pales ever more in comparison with the stakes of the Asian states themselves, led by economic superpower Japan. The rough magnitude of U.S. and Japanese investments, for the areas outside Japan where the United States and Japan compete, is clear in Table 8-2.

Looking to the future, there is every reason to suppose that the ongoing regionalization of trade and investment, both in Asia and the broader world, will continue. Indeed, it may well become even more pronounced. Canada, Latin

Table 8-2

U.S.-Japan Investments in Asia (1985-94)

	1985		1994	
	U.S.	Japan	U.S.	Japan
Indonesia	4.1	8.0	5.0	15.2
Hong Kong	3.1	2.8	10.5	12.7
Singapore	1.9	1.9	8.7	8.5
South Korea	0.8	1.5	3.0	4.9
Malaysia	1.2	0.6	1.9	5.6
Taiwan	0.8	1.3	3.1	3.7
Philippines	1.0	0.5	1.8	2.2
Thailand	1.0	0.4	2.9	6.5
China	—	1.0	0.9	6.2
Japan	9.1	—	31.4	—
TOTAL ASIA (outside Japan)	13.9	18.0	37.8	65.5

Sources: U.S. Department of Commerce, *Survey of Current Business*, August 1995; Keizai Kōhō Center, *Japan 1995: An International Comparison.*

Note: Figures are $ billions. U.S. figures in 1994 column are for December 31, 1993, while Japanese figures are for March 31, 1994, the end of Japan's 1993 fiscal year.

America, and the Caribbean provide American firms with tempting investment opportunities in energy and labor-intensive manufacturing, while Western Europeans can find similar prospects in Southern and Eastern Europe. Deepening political ties with neighboring nations may encourage favorable regulatory treatment where it is not already mandated by the expanding formal institutions of regional integration. And Westerners may find a transparency in their dealings close to home that they do not enjoy in confronting powerful, and culturally distinctive, Asian trading networks. The mutual bitterness provoked by confrontations like the 1994–95 U.S.–Japan auto-parts dispute does not help either.

With opportunity close to home, and with the macroeconomic handicap of low national savings, Europeans and Americans will find it hard to cut a powerful figure in Asia, relative to the increasingly affluent Asians themselves. The possible demise of Hong Kong as a globalized trading center, as it is absorbed by China, may compound their problems in the region. To be sure, with heavy current account surpluses contrasting sharply to European and American deficits, the

Asians will be in a position to venture actively beyond their own region, and will do so to some degree.[24] But with low labor costs, adequate technology, improving infrastructure, and a pool of increasingly affluent local consumers, their best opportunities will likely be close to home. Whatever Asians may do elsewhere in the world, it will be preeminently they who will build—or at least finance—the twenty-first century of their own region.

Will the American economic role in Asia be much more than window dressing, as the era of globalism recedes? As it declines, relative to the stakes of Japanese, Koreans, and overseas Chinese, how inclined will Americans be to provide a security umbrella in defense of Asian profits, especially should transpacific trade continue to be chronically unbalanced against them? Even if Americans are willing in the abstract, for the sake of globalism, to persist in their Asian security presence, how much will they be willing to pay to maintain it? How much, both in money and the currency of political concession, will Asians be willing to contribute in support of foreign forces continuing to linger on their own soil? Much turns on mutual conceptions of cultural and political community, not just the prevailing terms of economic interdependence, as the era of economic globalism recedes. These too are fundamental to understanding the challenge of Pacific Defense.

An Emerging Asian Identity?

Asians are growing closer economically. They are also, with their rising place in the world, growing more confident. As recently as 1960, Japan and East Asia together accounted for only 4 percent of world GNP, while North America made up 37 percent. Today the two regions have about the same share of global product (some 24 percent each). And Asia is growing faster.[25] Later in this decade, Asia's economy should be larger than that of the United States alone. Within a decade after that, China and Japan could *each* be larger, in the narrow scales of GNP calculation, than the United States.

Asia is also rising in technological terms. In a growing number of high-tech areas, such as amorphous metals and optical fiber research, Japan has already passed the United States.[26] With the post–Cold War assistance of impoverished Russian scientists, China and to some extent Korea are also gaining.

As Asia grows and becomes more self-sufficient economically, a parallel tendency toward intellectual and cultural self-sufficiency is emerging as well. Many East Asians are moving away from the assumption that what many view as a declining West should necessarily be setting standards for others whose societies self-evidently appear to be functioning more smoothly and efficiently. No longer is the mark of ultimate success simply admission to the Western industrial club.

Asians realize that it will be a struggle, as the prominent Singaporean civil servant Mahbutani puts it, "to work out social, political, and philosophical norms that best capture their peoples' aspirations."[27] It may be especially delicate and difficult to work out new concepts in sensitive areas like human rights, democracy, and freedom of the press. But many are determined to do so.

The impulse toward intensified Asian consciousness has more than simply economic origins. It is linked also to an erosion of local identities, provoked by economic modernization and increased mobility that is typical around the world.[28] National elites are also encouraging "Asian" consciousness as a way of insulating turbulent local populations from inconvenient foreign influences. In some countries it may also be a form of backlash against the preeminent position that the West, especially the United States, assumed with its triumph in the Gulf War, and with the collapse of Communism.

There is some danger that the new Asianism may further complicate differences of interest between Asia and the world that economic regionalism has already begun to set in motion. It provides the intellectual justification for all sorts of actions by East Asians that—in an age of instant communications—many Westerners both notice and often find troubling: Chinese human-rights violations and repudiation of Western intellectual-property protection standards; Japanese threats to use semiconductors as a trade weapon, coupled with blunt rejection of numerical measures for trade progress; and even things as mundane as Singapore's decision to cane an American teenager for spray-painting cars. Self-righteous, if sometimes chauvinistic, Western reactions are fed back across the Pacific, compounding the corrosive dynamic of cultural conflict that has begun to build, and that could ultimately threaten the pillars of transpacific cooperation so vital to Asia's own precarious security.

Neo-Isolationism in America?

East Asia—in cultural and psychological, as well as economic terms—is thus just not what it used to be. Like a newly liberated, restive marriage partner, it is ever more self-sufficient, with a firmer sense of itself and casting out in new directions. Many of the old links of the past continue to bind, however precariously. Yet shadows remain over the long-term future.

The redefinition of East Asia's sense of itself and its relationship to the world is important—and potentially threatening to the transpacific relationship as traditionally conceived. This is true precisely because of the myriad, complex, and asymmetric ways in which America and Asia, especially the non-Communist portion, are bound to one another. The United States has traditionally provided security and taken political initiative, while a poorer Asia, dependent on affluent American markets, has followed. Asia's changes have shaken the old assumptions and structure, posing the urgent yet open question of how America will react to its newly assertive transpacific partner.

As economic and cultural regionalism proceed, and as Asia becomes more critical and less deferential, how will these changes affect American political attitudes? How, in particular, will they shape U.S. conceptions of Asia, and American willingness to support existing national commitments there that are fundamental to Pacific Defense? The picture, after all, that we have painted here of Asia's future is one of disturbing hues. It is a canvas on which the balancing and liberalizing hand of the United States may well be crucial to both stability and prosperity, even as America's own relative economic presence in Asia continues to wane, and as local ambivalence about its broad geostrategic role begins to intensify.

Public opinion data suggest that 60–80 percent of the American people have consistently voiced support, or at least acceptance, of the necessity for active American involvement in international affairs, ever since the late 1940s.[29] In 1995, 65 percent of the American public favored this concept.[30] Yet within the parameters of the past there have been important nuances, of possibly fateful significance for the transpacific future. Three are especially worth noting here.

First, as one might expect from the underlying differences of economic interest between those directly engaged with the global economy and the general public, there are striking elite-mass contrasts in levels of support for an activist American role. In an important 1995 Chicago Council on Foreign Relations study, for example, "leaders" voiced virtually unanimous support (98 percent) for a U.S. global role, from the late 1970s into the mid-1990s.[31] Yet general mass public support for this concept has been much lower, and more volatile. From the 70 percent range in the 1950s it began dropping late in the Vietnam War, and fell to 54 percent—little more than half that of the elites—by 1982. The subsequent recovery has been slow and halting, to the 60 percent range, far lower than at the high point of the Cold War. Roughly a quarter of the American mass public, in most surveys, is generally considered isolationist, or roughly eight times the ratio of elites who feel that way.[32]

The isolationist sentiments of many in the general public come through clearly in answers to such questions as "What do you feel are the two or three biggest foreign policy problems facing the United States today?" Among the general public in the 1995 CCFR study, the most common three responses to this question were: (1) getting involved in affairs of other countries (19 percent); (2)

too much foreign aid (16 percent); and (3) illegal aliens (12 percent). Among "leaders," by contrast, the three greatest concerns were: (1) achieving free trade (24 percent); (2) dealings with Russia (23 percent); and (3) weak leadership (19 percent).[33]

A second broad pattern in public opinion surveys is the persistently fragmented nature of internationalist opinion in the United States regarding the use of American power overseas. It tends to be continually divided between a liberal "nonmilitary" internationalism of support for the United Nations, economic aid, and arms control on the one hand (perhaps a quarter of the population), and a more conservative "military" internationalism of support for a strong, interventionist defense posture (perhaps one-third of the total).[34] The shares vary at the margin, in pragmatic response to the success or failure in terms of human and material costs of concrete American interventions. Yet the division persists.

Within this general pattern of national ambivalence, there seems to be a declining willingness of the American general public to support international commitments of money and lives, as opposed to just rhetoric. In the 1995 CCFR poll, for example, defending the security of American allies was ranked "very important" by only 41 percent of the public, as opposed to 61 percent in 1990. Public support for the protection of weaker nations against foreign aggression fell from 57 to 24 percent.[35]

A third, related pattern in recent American public opinion is an increasing inclination to view national interests in economic rather than narrowly military terms. With the waning of the Cold War, this has meant greater mass sensitivity to the economic power of Asia, especially Japan and China, as well as Europe, and a greater primacy to economic policy among perceived national interests. In the 1995 CCFR study, for example, the public saw the economic power of Japan (62 percent) and China's development as a world power (57 percent) as more of a threat than Soviet military power (32 percent).[36]

These broad patterns, seen in conjunction, suggest that the U.S. security commitment to the defense of Asia—a pillar of Asian stability, as we have seen—may be dangerously vulnerable in domestic American politics over the long run. First, there is a substantial and persistent base of grassroots isolationist sentiment, although it still remains well short of majority standing. Second, dominant internationalist sentiment in the United States does not include a strong commitment to forward offshore deployment of military forces for its own sake. To the contrary, internationalists are painfully divided on the appropriate form of offshore American commitments, and increasingly disinclined to see the defense of allies as an overriding national priority. And finally, the increasingly economic cast of national security consciousness in the United States draws negative attention to the persistent and rising economic surpluses of Asia, particularly Japan and China, with the United States. These huge, intractable Asian surpluses, poorly justified in most American minds, make unilateral U.S. security commitments in Asia especially hard to sustain politically, in view of the rising skepticism about alliances in general.

In thinking about American political attitudes, it is crucial to distinguish what part of the country and what groups in society one is talking about. American attitudes toward Asia, like those toward world affairs in general, after all, have grown more and more schizoid over the past generation, as interdependence with the world has deepened. Since 1970, when we went decisively into trade deficit for the first time since World War II, America has grown ever more polarized between "territorialists," rooted in their local communities with little international involvement, on the one hand, and "symbolic analysts" whose interests and networks are global.[37]

The lion's share of the benefits to the United States from transpacific trade and investment over the past quarter century have accrued to the symbolic analysts—the investment bankers, attorneys, consultants, and executives of multinational firms in particular. It is they who have directly handled the working-level transactions through which economic interdependence has steadily deepened. For others in American society, the benefits of economic integration have often remained more abstract. Not surprisingly, working people without international ties have been more skeptical of the merits of interdependence, a pattern clearly reflected in the public opinion data presented above. Their enthusiasm for Pacific Defense, in all its manifestations, is clearly limited.

This basic picture of a bifurcated American politics, with a strong grassroots isolationist cast as the Cold War recedes, needs a bit of political-economic refinement. Actual interdependence with the world, and particularly with Asia, does, to be sure, generate some important, regionally concentrated economic interests that can critically influence the evolution of American policy at this twilight of globalism. For a while at least, they should help to sustain political collaboration otherwise threatened by skeptical currents of opinion.

Most crucial among these internationalist interests are agriculture, distribution, and construction, together with a few specialized, mostly high-technology manufacturers, who also hold powerful stakes in transpacific trade. These groups are concentrated in Sunbelt and Western states that became steadily more important in the American political economy across the 1970s and 1980s, and that are where the bulk of the three million American jobs generated by transpacific trade, as well as the eight million Americans of Asian origin, are concentrated. The most pivotal of these areas, the Golden State of California, with its more than thirty million people, casts fifty-four electoral votes—20 percent of the entire total needed to ensure a President's election.

U.S. trade with Asia has been chronically imbalanced for a generation, with the total U.S. transpacific trade deficit in 1994 substantially exceeding $100 billion. But the United States nevertheless managed a trade surplus of nearly $15 billion in agriculture with Asia, selling more than $8.7 billion in 1993 to Japan alone, $2 billion to Taiwan, $1.9 billion to South Korea, and $875 million to Hong Kong.[38]

This Asian market accounted for a full third of U.S. agricultural exports. Similarly, America recorded huge surpluses in aircraft, computers, software, farm equipment, wheat, coal, and lumber: East Asia was both the largest and most rapidly growing regional market in the world in almost every case. Boeing sold one out of every seven aircraft it produced to China alone.[39]

Manufactured inputs from industrial East Asia are also crucial to many American businesses, especially those in construction and distribution, which have powerful grassroots political networks. They are also important to U.S. multinationals, including the Big Three auto manufacturers, which handle large volumes of captive imports made in Asia, like the Dodge Colt.

Asian multinationals are also clearly important to the growing ranks of their American employees. Indeed, more than 500,000 Americans were working for Japanese firms alone in the mid-1990s. They included over 100,000 American blue- and white-collar workers in Japanese transplant auto factories, now producing over two million Toyota Camrys, Honda Accords, and Nissan Sentras along a five-hundred-mile stretch of Interstate 75 in America's industrial heartland, from Tennessee across Kentucky, Ohio, and Indiana into Michigan.

Confronting a $100-billion-plus total trade deficit with Asia, the United States clearly has groups being injured by transpacific trade, especially workers in autos, textiles, electronics, and related sectors. Reflecting the deindustrialization of America, the employment strength of manufacturing unions has been declining precipitously. The United Steel Workers, for example, has lost more than 500,000 members, or over half its strength, since the late 1970s alone.[40] The United Auto Workers has also been severely decimated.

Less than 16 percent of American salaried workers are now members of a union.[41] Declining membership rolls have obviously not enhanced the political influence of the unions. Their role was clearly less influential in the Democratic presidential campaign of 1992 than in either 1988 or 1984, despite Bill Clinton's victory. It was clearly ineffective in the Republican political landslide that followed, and seems unlikely to be restored even by the five-year merger of auto, steel, and machinist/aerospace unions announced in July 1995.[42]

A definite shift in the locus of regional political power since 1970, mirroring population shifts and long-term economic growth trends, has also rendered the U.S. economy more resistant to trade protectionism than was true a generation ago. The West Coast, the South, and the Southwest together have produced every U.S. President except one (the "accidental," unelected Gerald Ford) since John F. Kennedy's assassination in 1963. These regions are all growing in political stature. And across the broad swath of Sunbelt states from California to the Mason-Dixon line, there is not a single major steel mill, and only one auto plant— the NUMMI joint venture between GM and Toyota at Fremont, California. Even the semiconductor industry of California, with its specialization in microprocessors and software, competes only in limited fashion with the memory-oriented semiconductor industry of Japan.

All this adds up to a Middle America that does not take the lead in trade protectionism. Yet it lacks enthusiasm, with the Cold War waning, for America's persisting global security role. Even regions that are relatively liberal on trade matters are often frustrated at the grassroots over the costs that America bears for that global role, especially under conditions of slow economic growth where income inequalities are rising and tax burdens seem increasingly heavy and difficult to bear. The real wages of the average American, after all, have not risen in twenty years, which inevitably reduces any bias toward altruism that he or she might otherwise have.

If there is any common thread in American grassroots attitudes toward foreign policymaking since the collapse of the Berlin Wall, it may be a broad-based resentment at the perceived costs, and skepticism about the alleged benefits, of world leadership, especially of aiding or intervening in Third World nations of limited apparent relevance to American security. This has shown up most clearly in the backlash against foreign aid, manifest so strongly in the Republican 104th Congress.[43] It has also shown up in criticism of the United Nations, and American contributions to multilateral peacekeeping forces outside American control, that was so clearly expressed in the Contract with America.[44]

The Contract itself is clearly in the conservative "military internationalist" tradition discussed above, proposing expansions of American military spending and broader international commitments, especially through the expansion of NATO and greater activism against international terrorists. Yet there is a powerful, unilateralist, Fortress America streak, especially in relation to the United Nations, running through the document. The United States will either command or it will not get involved, except in very unusual circumstances. There is a parallel in the Contract to the psychology and politics of Super 301 trade legislation: use American power to get clear-cut results, and don't get bogged down in messy, untidy two-way bargaining. The emphasis on revival of the Strategic Defense Initiative antimissile defense has a similar unilateralist flavor.

Throughout the 1970s and the 1980s, the isolationist streak in America was largely populist and on the left. Now even the Republican Party, aggressively internationalist in the 1980s, has a clear isolationist wing. The libertarian Cato Institute has even argued that in the Persian Gulf, home of two-thirds of the world's oil reserves, there are no vital American interests.[45] It has also suggested withdrawal of all U.S. forces in Japan and Korea before the year 2000.[46] Ross Perot's 1992 presidential campaign and his struggle against NAFTA, like Pat Buchanan's appeals, show another dimension of neo-isolationism on the right.

Many recent opinion polls show a majority of the American people still half-heartedly supporting the use of American military power to defend allies in East Asia. In the abstract, we might be able to fight the Korean War again.[47] The problem comes in concrete willingness to pay actual costs, both material and human. With the flattening of American economic growth, rising income inequality, and the revolt against government spending in general, those forms of bottom-line resolve seem to be weakening. East Asian criticism of the American role and the American response to such criticism—amplified by a sensation-oriented mass media—will not make things easier.

The 1989 FSX controversy about sharing the costs and benefits of building Japan's next-generation fighter aircraft illustrates one sobering dimension of the emerging transpacific political dynamic at work. A development and production agreement was quietly concluded in the late 1980s by the Pentagon and the Japanese. Yet Congress, backed by the media, attacked it as a "giveaway." Congress adamantly demanded concessions, and the Japanese resentfully conceded, at least on paper.

The mechanical prototype of the aircraft was not delivered until January 1995, and it may well be close to the year 2000—at least two years behind schedule—that it will ultimately be deployed. Grassroots concerns for "fairness," filtered through Congress, were potent enough to force some formal Japanese concessions half a decade ago. But the working level in Japan quietly neutralized many of those apparent gains in the tortuous process of implementation. The overall loser, in this corrosive transpacific dynamic, was long-term cooperation, although no one was formally isolationist or uncooperative in any clear-cut way. Transpacific security relations are a fragile reed, and it does not take explicit isolationism or protectionism to damage them. Erratic political intervention and loss of mutual trust can also do the job.

Slow erosion of mutual confidence could be accelerated by sudden "trigger incidents" that rapidly intensify American isolationism.[48] Had the Gulf War not ended so rapidly and successfully, from a U.S. perspective, its human and material cost could possibly have provoked such sentiments. So could sudden incidents in East Asian seas or airspace, where U.S. forces suddenly incurred major casualties in defense of local interests, without adequate support from local allies. Korea could well provide such a case. Given the growing power of the mass media in American politics, such incidents could powerfully shape grassroots attitudes and make forward U.S. deployments in Asia and elsewhere even more difficult to sustain politically.

Overall, the American grassroots political impulse toward Asia seems to be status quo in trade policy, coupled with rising populist cries for retrenchment and expanded Asian "burden-sharing" on the security side, such as those surfacing in relation to foreign aid and the FSX controversy. The declining dominance of American firms in East Asian markets, relative to those of East Asians themselves, further compounds the skepticism of many Americans regarding future economic and political ties to East Asia. With respect to trade, aid, and investment, Japan seems to be steadily eclipsing the United States in Asia, while ethnic Koreans and Chinese of various national origins are ever more active across the region.

The danger is that U.S. corporate disillusionment with this deepening East Asian regionalist tendency could resonate with the populist American indictment of globalism more generally, and a backlash against increased East Asian assertiveness on transpacific issues, in a riptide provoking broader American disengagement from Asia. Such a development would compromise or remove a key stabilizing factor in the East Asian political equation that has, for a half century and more, inhibited the escalation of local intraregional tensions. In the end, it would fatally undermine Pacific Defense.

Fifty Years After: The Residue of History

America's economic presence in Asia may indeed be receding, at least in relative terms. Asians, more prosperous at home and less reliant on exports beyond their region, may be growing more critical of American ways. Yet this nation's security presence in the far Pacific endures so far, largely unchanged since MacArthur's days. The U.S. withdrawal from the Philippines notwithstanding, America's military profile in its western seas is receding far less rapidly than in Europe. Nearly 100,000 troops, together with a Navy carrier group, a Marine rapid-deployment brigade, and a substantial Air Force commitment, remain deployed in Japan and Korea.

Okinawa epitomizes, as suggested in Chapter 1, the slightly incongruous, antiquated realities of America's lingering presence in the western Pacific. Strategically located astride East China Sea maritime lanes, half an hour by F-16 from

Taiwan and less than an hour northeast of Shanghai, Okinawa is the natural cornerstone of a China-centric American Pacific defense deployment. More than 35,000 Americans serving with the Army, Navy, Air Force, and Marines are based there, as they have been since the bloody spring and summer of 1945. Indeed, the U.S. Military occupies more than 20 percent of all the land on the island.

In the shadow of its enduring American military presence, Okinawa retains a quaint 1950s flavor that has long since disappeared stateside. It still offers drive-ins, where waitresses in pinstripes serve root beer to motorists parked in their Toyotas outside. Less pleasant reminders for the local citizenry have long included a military port in the center of Naha, the local capital, that obstructs the main commuter route to downtown Naha, provoking monumental twice-daily traffic jams. Local politicians have pleaded with both Tokyo and Washington for years to close it, without success, since they can find no alternative locale willing to accept it. Periodically there are also incidents like the September 1995 rape case, which provoked the largest demonstrations in Okinawan history against the bases.

The U.S. military presence in Okinawa may seem but an exotic footnote in the grand scheme of history. Yet it combines complex symbolism and strategic significance in disturbing ways. Beyond the blue waters of the East China Sea looms a turbulent Asian continent with which both Japan and America have all too often been bitterly engaged.

As previous chapters have suggested, the forces for future insecurity in Asia remain powerful, despite the waning of the Cold War. Ironically, many of them, such as rising energy demand and related geostrategic rivalries, flow from the very Pacific prosperity that seems so reassuring at first glance. Economic globalism and the transpacific economic interdependence that has flowed from it have provided a firm basis for the U.S. military involvement in Asia that in turn mitigates such Asian insecurities.

Yet the economic equation, and with it broader, related interests, is definitely shifting. Economic ties between Japan and Asia are deeper now than they have been, while those east across the Pacific to America are weakening. Whether Okinawa as it now stands, in the globalism that it still symbolizes, can assure a pacific future for Asia beneath the Eagle's wings, as the shadows of regionalism and intraregional rivalry continue to deepen, remains to be seen. Therein lie major consequences for both the strategy and tactics of an effective Pacific Defense.

THE POLICY GAP

With the Cold War waning, young new leaders appearing, and the nations of East Asia moving rapidly to parity with America in economy and technology, a new world is clearly emerging in the Pacific. It creates new imperatives for Pacific Defense. Asia and the United States will need to look at each other afresh as their relationships inevitably change. Yet their policies toward one another—and even their perceptions—are profoundly rooted in the past. Both sides are prisoners of institutions created in a world long gone, to fill once-pressing needs that no longer exist.

The key nations of the Pacific were more similar in many key respects—and more flexible in dealing with their underlying problems—on the eve of the Great Crash in 1929 than they are today. Bureaucracies were less interventionist and trade barriers were less elaborate than they later became. Indeed, it was the turbulent Trans-War Era (1930–55) of response to depression and armed conflict that gave the public institutions of the Pacific their present, in many ways problematic, cast. The diversity of institutional forms born in that fateful Trans-War Era lies at the heart of the growing tensions across the Pacific that Asia and America sadly but inevitably confront today.

In the United States, the trans-war decades brought first depression and then

CHAPTER 9

global warfare, creating a set of national institutions with dual preoccupations: New Deal–style welfare and national defense. Obscured in this mixture was what had previously been a high priority—forthright government support for domestic business interests. The Commerce Department, which had been the most powerful part of the Executive Branch in the 1920s under Herbert Hoover, was eclipsed and emasculated, transformed into a hodgepodge of miscellaneous, often unrelated fiefdoms.

In its place at the center of national policy concerns arose the CCC, the NPA, the Social Security Administration, and a blizzard of other new welfare institutions, together with a consolidation of the military services in a cavernous new wartime Pentagon. Crucially, the previous flexible communication between government and the non-defense-related business sector was also severely compromised by new adversarial sentiments and legalism on both sides. By 1950, with the CIA, the NSC, the CEA, the Defense Department, and other institutions of a powerful new, security-oriented Imperial Presidency in place, the new defense- and welfare-oriented structure of American policy formation was largely complete.

Japan began remarkably close to America's own starting point—a powerful private sector and a limited central government, supportive of business. But it ended up at a very different destination. Japan's formative experiences—total mobilization for war, disastrous defeat, unconditional surrender to a powerful adversary, and then occupation by a foreign army—were more traumatic than even the wrenching pressures that America had undergone. Japan's economic institutions were transformed much more profoundly than America's, allowing a sharply enhanced role for state direction.

In the aftermath of defeat, Japan's political institutions also underwent much more sweeping change. The occupation dismantled both the military and the Home Ministry, which had earlier rivaled and even eclipsed the economic bureaucracies in its policy influence. The outcome was a Japanese state singlemindedly pursuing national economic prosperity, with export promotion as a central tool of policy. That was precisely what Washington, fed up with subsidizing the chronic trade deficits of a sickly Japanese economy with American aid, itself desired.

China, unlike the United States and Japan, was shaped by a truly revolutionary past—particularly the formative experiences of the Long March (1934–35), the Civil War (1946–49), and the Communist consolidation of power that followed thereafter. These experiences, to be sure, created institutions, such as the People's Liberation Army and the Chinese Communist Party itself. Those continue to profoundly shape China's orientation toward the world to this day.

But almost as important as the institutions themselves was what history did not create in postrevolutionary China. It did not, in contrast to the Japanese pattern, create or reinforce a powerful bureaucracy, oriented toward consistent, if rigid and uninspiring, behavior. To be sure, such bureaucracy had long been a Chinese tradition, and its residual power has been a persistent feature of the Chinese political landscape. But it was precisely Mao's fervent, iconoclastic desire to destroy or at least measurably weaken this powerful conservative force, a task in which he was marginally successful.

The Revolution, with its populist dictates ("It is justified to rebel"; "The army is a fish in the sea of the people") created a decentralized, nominally grassroots-oriented state apparatus, building on Mao's decentralist bias. That structure has been, despite the radical heritage of the Great Leap Forward and the Cultural Revolution, remarkably pragmatic. Yet the Chinese state has also often proved distressingly inconsistent and destabilizing in its international conduct.

Ironically, that Chinese revolutionary state also became one in which, because of inevitable fragmentation of power in a continental land of over a billion people, connections (*guanxi*) were crucial. The outcome has been a schizoid Chinese state, welcoming foreign investment while simultaneously sponsoring pirate software factories, exporting missiles, and extending nuclear technology to rogue nations for the private profit of well-placed yet nominally proletarian officials and their associates.

The policy "machines" of other Pacific nations have similarly been shaped by history, in the general mold of one of the three archetypes: America, Japan, and China. Australia and New Zealand, broadly speaking, have a security and welfare orientation analogous to that of the United States embedded in their institutions. North Korea, Vietnam, and to a large extent Russia still retain the dubious heritage of socialism: a combination of decaying, corrupt state institutions, together with an intricate web of interpersonal contacts to manage them. That pattern is characteristic of China, albeit perhaps with more of a centralist bias than massive, nominally populist China itself can afford to maintain. South Korea, Taiwan, and Singapore conform to the Japanese pattern of a powerful state economic bureaucracy, although their militaries, especially in the first two cases, are substantially stronger in political terms than is that of Japan.

History has fatefully shaped not only the domestic policy "machines" of Pacific nations, but also the framework within which they deal with one another. Most important, it created the stark Cold War political-economic divides that estranged the continental Communist powers (China, Russia, Vietnam, and North Korea) from the rapidly growing Pacific Rim nations allied with the United States.

Within the Pacific Rim, history and its Yankee co-conspirators created the San Francisco System, linking the United States with Japan, South Korea, Taiwan, the Philippines, Thailand, and ANZUS (Australia and New Zealand together) through a complementary set of bilateral economic and security relationships carefully designed to contain China.

Under this framework, created by John Foster Dulles in the process of concluding the San Francisco Peace Treaty of 1951 that formally ended World War II in the Pacific, the United States proposed bilateral security guarantees. These often included the local deployment of U.S. forces, coupled with broad trade access to the wealthy U.S. market to sweeten the deal. Over time, particularly since 1993 and the birth of APEC summits, America has begun to pursue broader multilateral frameworks, linking nations of the region with one another as well as with the United States. Yet the fundamental heritage of history for Pacific institutions has contrasted sharply to those of Europe. Asia has no NATO, and no analogue to the European Union: instead, it has an interlocking system of hub and spokes—with the Asian spokes very decidedly revolving around the powerful hub of Washington.

The Potential of the Private Sector

Complementing the outmoded and incomplete public institutions of the Pacific, fortunately, is a sophisticated and dynamic private sector. Like its public-sector counterpart, the private sector of the emerging Pacific Basin is highly diverse, reflecting the contrasting histories of its member nations and cultures. Yet its various types of private firms, industrial groups, and trade associations are linked by a common commitment to efficiency, profit, and innovation that make them a useful supplement or alternative to the more parochial and rigid hand of the outmoded states that populate the region.

Most elemental, perhaps, are the myriad family firms, mostly small and ethnically Chinese, that provide the backbone of economic life across Southeast Asia.[1] They lack complex organizations, but operate quietly and effectively in even turbulent and uncertain environments. Handling everything from gold and drugs to hand soap and perfume, while running hotels, restaurants, and mah-jongg parlors, they assure that consumers are supplied with what they want across national boundaries, even in the face of complex formal trade restrictions.

Larger-scale are the general trading firms (sōgō shōsha) of Japan, and their counterparts in Korea. These hybrid firms, which range broadly across the region, handle imports, exports, internal domestic trade, resource development, and even technology transfer. Most of them are closely linked to large commercial bands and manufacturers to form industrial groups (*keiretsu* to the Japanese and *chaebol* to the Koreans).[2]

The Pacific has also, in recent years, become a major base of operations to myriad Western-style multinational corporations. Most venerable are the banks, oil companies, and, surprisingly, automobile manufacturers like General Motors and Ford, which began operating actively across national boundaries in the region before World War II. After setbacks at the hand of local government for diverse reasons during the 1940s and 1950s, Western multinationals grew highly active again during the 1960s, supplemented gradually by Japanese, Korean, and Southeast Asian counterparts as those nations have become affluent capital exporters over the past generation.

Among the varied private sectors of the Pacific nations, those of East Asia have most clearly grasped the long-range potential of transpacific interdependence. It is easy to see why, confronting a large, affluent America—based in a less affluent Asia, and strongly supported by their own local governments in many cases—they acted as they did.

What is more paradoxical is the mixed behavior of the American private sector. Some of its firms, such as IBM, General Electric, Boeing, and several major oil companies, have been remarkably farsighted and dynamic in grasping and exploiting the potential of the Pacific. Yet all too many—including large numbers that have shown great creativity elsewhere in the world—have not emerged decisively in Asia.

For many years a major problem for American firms was the malady plaguing the automobile industry—a domination of key manufacturing firms by lawyers and accountants not concerned much with manufacturing competitiveness.[3] Many U.S. companies, as profit maximizers, were also mesmerized by lucrative, stable opportunities in the defense industry. For all too long it was more profitable for Chrysler to make M-60 tanks than to compete with the Japanese in compact cars.

The lure of defense was clearly another major factor inhibiting U.S. electronics and machine-tool manufacturers from responding more aggressively to East Asian inroads in their sectors. But to say that U.S. private firms were distracted from responding to East Asia is not to deny their inherent and demonstrable dynamism and creativity. The central problems of America's response to the Pacific, and the resulting gaps in Pacific Defense broadly conceived, lie elsewhere.

The Failures of America's Policy Machine

From the outset one must recognize a failing that has plagued American East Asia policymaking in virtually all policy sectors for many years: an inadequate understanding of East Asian history, culture, psychology, and values among the people who actually make U.S. policy. Robert McNamara recognized this failing as being crucial to mistakes that were persistently made by U.S. policymakers during the Vietnam War, such as systematically misunderstanding Chinese intentions and Vietnamese nationalist aspirations.[4] A parallel judgment could be made about failures in economic policy toward Japan early in the Clinton administration, such as those leading to the breakdown of the 1994 Hosokawa-Clinton summit, or in nuclear negotiations with the North Koreans later the same year.[5]

The remedy is not simple. It clearly cannot be a simple takeover of American Asia policy by Asia specialists, who after all have a parochial bias of their own. Yet their specialized background needs systematically to be factored into broader policymaking. Among various organizational forms, a "matrix organization" combining functional with area-studies expertise—like the State Department—may be especially effective, if it includes politically influential, culturally sensitive individuals specifically tasked with this integration. This general problem of the policy gap deserves more detailed treatment in various specific policy contexts, which we will try to give it later.

The most enduring defect of America's policy machine, as it confronts a rising East Asia, is its myopia on economic matters. The U.S. government, despite a small but excellent corps of specialists on Asian economies, has few good institutional mechanisms for detecting, assessing, interpreting, and responding to economic developments across the Pacific. Until quite recently, it has also lacked effective communications with the American private sector on such crucial questions. As a result, this country has repeatedly failed to even detect embryonic long-run trends in the region. The rise of Japan in the 1950s and 1960s, the hyper-growth of South Korea and Taiwan from the mid-1960s on, and the steady expansion of China since 1978 are only a few of the epochal Asian economic developments that came appallingly late to our national radar screen.

To be sure, the very nature of East Asia's rise—gradual, cumulative, and hardly dramatic—has made detection difficult. It is hard to point, even in retrospect, at many real watersheds. Only those with a long-term perspective and very low metabolism would be prone to even detect the emergence of an East Asian edge, let alone react with any dispatch. Yet on matters of such momentous national importance it is sobering that neither government nor government and business together did better than the historical record suggests.

Part of the problem stems from the *modus operandi* of American government itself. In sharp contrast to East Asia, the top ranks of the American bureaucracy are all political appointees, who normally serve in their government posts a bare two years, or even less. Neither these leaders nor subordinates, responding to their priorities, have strong incentive to think and act for the long term. And in a spartan world of sharp budget cutbacks, few feel they have the luxury of devoting scarce resources to the abstractions of the distant future in any case. This dangerous bureaucratic bias is compounded by the persistent tendency in Congress, under both Republican and Democratic leadership, to reward projects with high short-term visibility, such as "diplomatic security" against terrorists, as opposed to goals, like systematic cross-training for security and economic specialists, that offer only longer-term payback.

Paralleling the contrasts in future orientation within the American and East Asian economic bureaucracies are analogous differences in attitudes toward the past. East Asian government bureaucracies, like their private-sector counterparts, consciously cultivate institutional memory and try to learn systematically from what has gone before. Americans burn or shred many institutional files when administrations change, leaving newcomers remarkably naive about what has already been tried, even by their immediate predecessors.

Japan's MITI, in contrast, publishes a new institutional history every five years or so. Indeed, it maintains an office specifically tasked with keeping historical track of ministerial decision-making. America's Commerce Department, a full quarter century older than MITI, has never published a sound, comprehensive history in its nearly one hundred years of existence.[6]

America's problems of public myopia in relation to East Asia do not flow mainly from inadequate staff resources; Japan's MITI has only twelve thousand career employees, for example, yet is arguably much more efficient and strategic in promoting national economic interests than a sprawling U.S. Commerce Department three times its size. Yet there are two decisive transpacific differences that favor the East Asian bureaucracies: internal resource allocation and the quality of government-business communication. East Asians tend to favor long-term planning, institutional memory, and other organizational functions with consequence for the future. They also, in contrast to common, albeit changing, American practice, encourage active business-government dialogue.

Such communication tends to be more candid, and higher in quality, in East Asia for three reasons. First, the private sectors of East Asian nations themselves tend to be more organized than their counterparts in the United States, a few exceptions like semiconductors notwithstanding.[7] Organized private business, backed by formal industry associations, for example, can more easily support analysis of the long-term future than a structure that is more individualistic. Second, both private and public representatives in East Asia are, as a consequence of lifetime employment, more rooted in their organizations than is common in the United States. They are hence more prone to share mutual, institutionally related confidences.

Organizational structure is certainly different in U.S. and East Asian government, although the differences are a mixed blessing for both sides. The United States, for example, lacks a department of trade and industry, in contrast to virtually every other industrialized nation. Such a body could systematically provide the long-term analysis and detailed understanding of specific sectors, together with the political support for trade negotiators, that Japan's MITI and analogous bodies in other industrialized nations do. It could also, however, be a magnet in the American system for inefficient protectionist lobbies and an inhibitor of innovation, as bloated socialist bureaucracies, and at times the U.S. Commerce Department, have often been.

The United States, again in contrast to most G-7 nations, not to mention most Asian states, also lacks a department of construction. In both Japan and South Korea, by contrast, the construction ministries are among the most powerful institutions in government. They play a key role in domestic civilian construction, research into advanced construction techniques, and support for domestic construction firms in international competition. Yet they are also a forum for political corruption.

The closest approximation in the United States is the U.S. Army Corps of Engineers. Since 1824 the Corps has steadily expanded its civilian public works responsibilities, with vigorous support from Congress, to include planning and building of dams, canals, roads, railroads, and bridges. But virtually all the Corps's work has been domestic, some work in the Saudi Arabia of the late 1970s being a major exception. With more than $500 billion in public works spending planned over the next decade, East Asia, like the Middle East of the early 1970s, stands on the verge of an infrastructure construction boom. The U.S. government does not seem configured either to comprehend or to exploit this major new emerging opportunity, although the Commerce Department has recently made some major effort.

Telecommunications is another area where U.S. governmental structure diverges from that of most of America's economic competitors. Its important and rapidly expanding telecommunications sector is rather highly supervised by a small independent agency, the Federal Communications Commission, and several federal court decrees. Most nations have a full-fledged telecommunications ministry, and facilities traditionally owned by a major public corporation.

Overall the United States is probably best off in economic competition with something like its existing market-oriented, lightly supervised pattern of government. Certainly the American approach generates clear competitive advantages in dynamic, market-driven sectors like telecommunications that are likely to become more important over the coming decades. To be sure, there is something to be gained by "reinventing government": making it more decentralized, more flexible, and more communicative with the private sector.[8] America can learn, in a limited way, from East Asia in this regard, particularly in terms of incorporating private-sector views systematically into policy and maintaining a national focus on long-term issues. But the dangers of overregulation—in Asia as well as Europe and America—are becoming clearer with every passing day.

As serious as the U.S. government's failure to respond to long-range economic trends are its problems in handling relationships between economics and security. These interrelationships are enormously important in the Pacific and fundamental to effective Pacific Defense. Indeed, as we have seen in this book, many of the historic economic changes now under way there, such as Asia's growing economic scale, its emergence as a global technology center, its chronic energy shortages, and its rising energy links to the Middle East, have profound security implications. So do the turbulent ongoing economic and social transitions of China and the Northeast Asian Arc of Crisis, which in turn themselves bear potentially serious implications for both Japan and the U.S.–Japan relationship. America will need economic as well as conventional military tools, including support for cooperative energy, food, and environmental programs, to respond to these historic changes.

There are, to be sure, isolated parts of the federal government that try to cobble together concerns for both economics and security. Among the most important, perhaps, are the efforts within the State Department itself, whose enormous fund of expertise on Asia is one of the great inadequately tapped resources of the government as a whole. U.S. embassies abroad, including the crucial post in Tokyo, are increasingly cross-training their diplomats between economic and political sections, to develop an integrated appreciation of the economics and security interface.

There are still integration problems in Washington, both between the functional and regional bureaus and between economics and security-oriented functional bureaus themselves.[9] In particular the functional bureaus seem not to be adequately using the sometimes subjective yet culturally sensitive insights of the regional bureaus,[10] as was clearly true in the formulation of policy toward North Korea during much of 1994. Yet the potential power of "matrix organizations" like the State Department to mobilize detailed local reporting in an analytically coherent economics and security framework is so great that internal coordination problems within such agencies need priority attention.

The White House, of course, is another place where this integration of economics and security thinking needs to be achieved. The President has, indeed, a Special Assistant for National Security Affairs and Economic Policy in the White House. That key official has overlapping relationships with both the National Security Council and the newly formed National Economic Council.

Yet in the mid-1990s, more than a generation after the oil shocks and the rise of East Asia were fully in view, this special assistant remains only one of fifty-six senior officials at that rank in the White House, with vast, almost unmanageable global responsibilities at his vital interface, and limited staff support. In the intelligence community and the Pentagon, concern for the economic dimension of national security, including East Asian aspects, is clearly rising. Yet the institutional heritage of past separation between security and economic concerns still makes an intellectual integration painfully difficult to achieve.

Rapidly changing technology compounds the problem. High-resolution satellite photography, advanced eavesdropping techniques, and high-speed, computerized signal processing have given U.S. intelligence, particularly the National Security Agency, very sophisticated ability to monitor "hard" intelligence on military capabilities and troop deployments around the world.[11] Such remote, automated signal processing, operating from the depths of space, was remarkably good at picking up ominous early signs of North Korea's fledgling nuclear program at Yongbyon—indeed much better than the North Koreans probably dreamed it could be. Satellite imagery painstakingly tracked Yongbyon's evolution from a small two-to-four-megawatt reactor, sold by the Russians to North Korea in 1965, to the heart of a major nuclear complex. It showed in detail a research installation, roads, railroad tracks, power grid, housing, and storage areas, as the complex was slowly unfolding.[12] It starkly revealed the ominous lack of power transmission lines at Yongbyon that suggested its standing as a military rather than a civilian-power-oriented reactor. Yet Pyongyang had no way to know the degree of clarity with which U.S. intelligence was quietly observing its operations.

With their undoubted technical strengths, signals intelligence and aerial photography have attracted a growing share of U.S. intelligence funding in recent years. Yet they have inherent shortcomings. They simply cannot deal with the complex analytical questions at the interface of economics and security that will be increasingly crucial to America's future in the Pacific. And they can be misused.

Nowhere are the institutional failures of our current policy machine—mindlessly churning out yesterday's solutions to tomorrow's problems—more glaring than in America's approach to energy policy and its security implications. As we have seen, energy shortages are East Asia's greatest single vulnerability—its Achilles' heel. Energy shortages are driving East Asia steadily toward nuclear power, deepening tensions over offshore oil prospects, and inspiring naval rivalry over energy lifelines to the Persian Gulf. Yet these troubling long-term security questions—profoundly linked to economic change—are virtually off the radar screen of the U.S. government. The institutional failures of our policy machine bear much of the blame.

Since 1977 the United States has had a Department of Energy (DOE), with around twenty thousand regular employees. It is nearly double the size of Japan's MITI, even before DOE's 150,000 contract workers are figured in.[13] In 1994 it spent $17.5 billion a year. But DOE accords remarkably little priority to the important emerging energy issues in the Pacific, and it lacks in any case the institutional power to galvanize a dynamic response. This myopia and decision-making paralysis are glaringly ironic: DOE was a child of the oil shocks, conceived precisely to enhance American ability to foresee and respond to the linked problems of economics and international security that those shocks so starkly posed.

Despite the circumstances of its birth, DOE also has a Cold War heritage that complicates its response today to emerging issues. DOE's most important institutional forebear is the Atomic Energy Commission, from which it inherited responsibility for the manufacture of all nuclear weapons in the United States, and their safe delivery to the Pentagon. Even though America has not produced a single nuclear bomb since 1990, bureaucratic wheels in the strategic nuclear area continue to churn. Close to 70 percent of DOE employees and budget are still devoted to either cleaning old nuclear weapons plants or maintaining residual production capacity for the future—functions that cannot be ignored, but that hardly justify their current priority in men and material.

Apart from supervising nuclear weapons manufacture, DOE's most important alternative mission in the 1970s and early 1980s was enforcing price controls over oil and natural gas, in the wake of the oil shocks. With the collapse of oil prices and the move to energy deregulation, such responsibilities, together with the political leverage that accompanied them, also disappeared. DOE is now left with a potpourri of miscellaneous, often mismatching responsibilities: regulating the coal and oil industries, statistical data collection, supervision of twenty-eight national laboratories, management of strategic petroleum reserves, and developing alternative energy sources (but not supervision of auto fuel-economy standards).[14] This grab bag confers little ability to coordinate energy policy as a whole. And the considerable headaches of complex day-to-day administration leave top officials with little time to think long-term, toward important emerging issues such as international energy cooperation in the Pacific.

World War II and the National Security Act of 1947 created a muscular, coherent set of defense organizations in the Executive Branch of the U.S. government—the National Security Council, the Central Intelligence Agency, and a consolidated Department of Defense. This defense establishment remains, to this day, the most distinctive, and probably the most powerful, institutional feature of the U.S. government. Yet its members too, like the Energy Department, find it difficult to focus systematically on how to link economics and security. And even when they identify the problem—in the Pacific energy area or elsewhere—the optimal solution often lies beyond their sphere of operating responsibility.

Americans may not object, for example, to minor expansions of Pentagon operating authority to resolve clear security threats. Even a heavily Republican Congress acceded, for example, to a Democratic President's late-1994 proposal to release Defense Department emergency petroleum reserve funds to purchase heavy fuel oil that would help induce North Korea to abandon its menacing nuclear program.[15] But most Americans would still prefer to see broad international cooperative energy programs in other hands than those of the Pentagon, and to see the Pentagon focus on more conventionally defined national-security activities. And there are no other government—or indeed private—hands to actively manage such undertakings, other than possibly a State Department that, for a variety of bureaucratic and domestic political reasons, is frequently discredited.

Looking to the future, America must anticipate an ever more interdependent, technically sophisticated, and rapidly changing world, in which communication is virtually instantaneous. In this world of the Third Wave,[16] bureaucracies—unless decentralized and responsive to their changing environments—will be increasingly obsolete. Conversely, strategically conceived interpersonal networks will be ever more important to both understanding and action.

Given the personalism of East Asian political and economic systems—the visceral need across most of the Pacific for person-to-person communication on delicate matters—there will be no substitute for grassroots diplomacy and intelligence-gathering. Embassies that can combine first-rate reporting and analysis will not be obsolete. Yet America will need to place a new premium on the software of human relations, as opposed to the "hardware" of institutions. It needs a special focus on the strategic creation of networks and forums for improving the flow of information and ideas relevant to Asia policy—both among Americans and between American and Asian counterparts. For it is there that the most serious policy gap for America will lie.

Meaningful networks are easiest to form, in the East Asian context, just outside the government. There an understanding of actual policy dynamics coincides with the flexibility of freedom from the legal and operational constraints operative in the public sector. Unfortunately for American interests, East Asians took the initiative early in this shadowy world, developing an intricate web of Washington lobbyists and consultants—many former U.S. government officials—that had few parallels in Tokyo, Seoul, Taipei, or elsewhere in Asia.[17]

During the early postwar period a few Americans were very sophisticated in networking and lobbying in East Asia.[18] They demonstrated that—contrary to revisionist rhetoric—such activities were possible. Yet over the 1960s and 1970s, such skills clearly atrophied. They need to be revived, not just for a few firms, as is currently the case, but for all American industry and government, in this nation's deepening relations with Asia. They are a basic, "forward-deployed" element of Pacific Defense.

Japan's Dangerous Policy Gap

Across the Pacific, as here in America, Japanese government agencies and private firms are also configured to deal with the problems of a comfortable past unlikely to resemble those of a more ominous future. They too churn out policies and programs unrelated to the times. But Japan's policy gap is the reverse of that of the United States, centering on security rather than economics.

Seen in comparative perspective, one of the most striking realities of Japan's public foreign policy institutions is their small scale. With only 4,400 employees (60 percent of them serving abroad), the Japanese Ministry of Foreign Affairs (MOFA) is smaller than counterparts in Canada and Italy, one-half the size of Britain's, and less than one-third the size of the U.S. State Department.[19] Japan's Ministry of International Trade and Industry (MITI), which combines a broad range of strategic functions with respect to trade, industrial, commercial, small business, and technology policy, has only 12,000 employees, or roughly one-third the staff of the U.S. Commerce Department.[20]

Japan's foreign aid bureaucracy is also remarkably small. In 1988, for example, Japan had only 1,539 personnel serving in its two foreign-aid disbursement institutions (the Overseas Economic Cooperation Fund and the Japan International Cooperation Agency), compared to the 4,695 who disburse a roughly equivalent amount of American aid. Japan's foreign aid disbursement staff has grown less than one-third as fast as the overseas development assistance (ODA) budget, year in and year out, ever since the mid-1970s.[21]

Despite the major expansion in Japan's global economic activities over the last decade, and a doubling in its share of world GNP to 15 percent of the global total since 1985, Japan's foreign policy bureaucracy has hardly grown at all in size, even from its modest prior levels. In 1985 the Japanese Foreign Ministry had 3,883 employees around the world, distributed among 110 embassies and sixty-two consulates general abroad, in addition to various responsibilities at home. By 1993, following the dissolution of the Soviet Union and a proliferation of diplomatic responsibilities in the developing world, Japan had 181 embassies and even more consulates than before. Yet despite an increase in diplomatic outposts abroad of nearly a third, the number of personnel in the Ministry of Foreign Affairs had risen only 16 percent, to 4,636.[22]

Meanwhile, at MITI the number of career personnel was actually declining—from 12,672 in 1985 to 12,376 in 1993.[23] Despite an increase in the scale of Japan's involvement in the international economy along the trade, investment, and technological dimensions, deregulation and budgetary constraints imposed by the Ministry of Finance (MOF) systematically impeded MITI's expansion to fill new global roles. Balance with domestic ministries (whose staffs fell by nearly 2 percent over the 1985–93 period, as part of a determined effort to cut budget deficits) took priority in MOF's mind over any recognition of new global responsibilities.[24] The problem was intensified in 1992 when the Diet rejected Prime Minister Miyazawa Kiichi's proposal for an "international contributions tax," in the aftermath of the unpopular Gulf War.

Japan's presence in multilateral public institutions is likewise small and static. As is well known, Japan continues to lack permanent membership on the United Nations Security Council, even though the United States and an increasing number of European permanent members of the council voice at least nominal support for such standing. Only about one United Nations staff member in thirty is now Japanese, although Japan pays four times that share of the UN's annual budget.[25] Britain, by contrast, in 1994 had almost exactly the same number of UN staffers as Japan, but contributed only 40 percent of Japan's share of the UN budget. Indeed, every one of the major Western nations had more UN staff appointments, proportionate to contributions, than did Japan.

To be sure, a few individual Japanese are coming to play leading roles at the top levels of multilateral institutions. Akashi Yasushi, for example, served during 1994–95 as chief UN representative in the former Yugoslavia, after heading the successful 1992–93 UNTAC effort to bring peace and free elections to Cambodia. Ogata Sadako headed the UN High Commission on Refugees. Yet these prominent Japanese are few and far between. Far more common is the pattern at the International Monetary Fund, where Japan has consistently, across the 1980s and 1990s, declined to challenge French incumbents for the executive directorship of the fund, despite its unambiguous standing as by far the world's most substantial creditor nation.

The most conspicuous institutional weaknesses of the Japanese state are clearly in the political-military area. The Japanese Self-Defense Forces themselves, operating under the constraints of a 1947 "no-war" constitution that proscribes offensive military capabilities, are relatively small by international standards, although mobile and well armed. In fiscal 1993, total Japan Defense Agency (JDA) personnel, both civilian and military, totaled 299,274, or fewer than the Ministry of Posts and Telecommunications, with 307,662.[26] Japan's actual armed forces personnel strength of less than 240,000 was dwarfed by that of its neighbors: 2.9 million men under arms in mainland China, 1.1 million in North Korea, 633,000 in South Korea, and 376,000 in Taiwan.[27]

Although the JDA now disburses a formidable $50 billion plus defense budget, it remains highly constrained within the institutional framework of Japanese policymaking as a whole. Quite apart from Article 9 of the constitution, by which Japan forswore the right of belligerency, the JDA is limited by its own inferior institutional standing within the Japanese bureaucratic hierarchy. While nominally a member of the cabinet, the JDA director-general is not a full-fledged minister of defense, but rather a second-ranking "minister of state" (*kokumu daijin*), equivalent in standing to the head of the Economic Planning Agency or Environmental Protection Agency.

Within the JDA itself, strategic divisions are dominated by other ministries, which orient Japanese defense policy systematically in accordance with their own institutional interests. The Accounting Bureau, for example, which is charged with preparing the defense budget at its most sensitive early stages, is traditionally headed by an official of the Ministry of Finance on secondment. This MOF professional within the JDA coordinates closely with budget examiners within MOF's own Budget Bureau, preempting efforts by the uniformed services to inflate their claims. This institutionally based intrusion of conservative MOF officials into internal JDA processes—highly unusual in comparative perspective—has crucially inhibited expansion of the Japanese defense budget for more than three decades.[28] Similarly, the traditional appointment of a career MITI official as director of the JDA Procurements Bureau—the responsible official negotiating with the Pentagon and other foreign authorities on procurement of military equipment by Japan—has tended to subordinate issues of strategy in equipment acquisition to industrial-policy considerations.

Japanese intelligence capabilities, especially in the political-military area, are also weak. This deficiency has significantly intensified the hesitancy of Japanese crisis decision-making on matters far from Japanese shores, a tendency that showed up graphically in the Persian Gulf crisis of 1990–91.[29] There is no clear institutional analogue in Japan to either the U.S. Central Intelligence Agency or to the Pentagon's Defense Intelligence Agency; only in the late 1980s did the Ministry of Foreign Affairs (MOFA) establish an autonomous Information and Research Bureau (Jōhō Chōsa Kyoku) roughly analogous to the U.S. State Department's Bureau of Intelligence and Research (INR). The primary intelligence-processing entity directly advising Japan's prime minister, namely the Cabinet Information Research Office, is less than one-hundredth the size of America's Central Intelligence Agency.

Even in the aftermath of the Gulf War, when the deficiencies of Japanese intelligence were so starkly evident, the Information and Research Bureau remained by a significant margin the smallest of MOFA's nine bureaus, with a staff of only fifty-eight.[30] This compared to a staff of 349 in MITI's economically oriented Research and Statistics Section. MITI's unit, in contrast to the MOFA pattern, was directly attached to the Ministerial Secretariat and was further augmented by an internal Research Institute for International Trade and Industry (RITI), headed by the distinguished former Tokyo University economics professor Komiya Ryūtarō.[31] Both MOF and the Bank of Japan likewise had in-house analytical capabilities considerably stronger than MOFA's.

Apart from MOFA's Information and Research Bureau, the only other actual public intelligence-gathering organizations in Japan are the Public Security Investigation Agency (PSIA), affiliated with the Justice Ministry; the Ground Self-Defense Force Staff Officer Annex (Nibetsu); and the foreign affairs and security sections of the National Policy Agency. The Public Security Investigation Agency amasses and analyzes information concerning groups within Japan, particularly those on the extreme right and extreme left, that are deemed potential security threats, although Japan continues to lack a clear-cut espionage law with enforceable penalties. Because of the presence of perhaps 430,000 ethnic Koreans with North Korean sympathies within Japan and the importance to North Korea's economy of the roughly $1 billion flowing annually from them to the North, the PSIA has significant insights into at least one issue of major international importance. Nibetsu monitors radio transmissions, especially those of the former Soviet Union, China, and North Korea. Although its technical signal-intelligence capacities are limited, the quality of its work is considered to be high, as evidenced by the Japanese radio intercepts of Soviet military communications revealed publicly at the United Nations after the KAL 007 airliner was shot down by the Soviets off Sakhalin in the early 1980s.

The foreign affairs and national security sections of the National Police Agency also have some important, if limited, intelligence-gathering functions. The foreign affairs section concentrates on anti-Japan operations carried out by Russia, China, North Korea, and other reputedly hostile countries, while the security section monitors the movements of political activists such as the Japanese Red Army.[32] The operating roles of both sections were profoundly shaped by the Cold War and the rash of hijackings of Japanese aircraft by the Red Army in the 1970s; they now appear curiously anachronistic, in the minds of many, following the collapse of the Soviet Union and the recent Israeli-Palestinian peace agreement.

Japan has, in summary, some significant, if limited, intelligence capabilities with regard to events around its own periphery, most crucially economic and social conditions in North Korea. Indeed, its insights into the nonmilitary aspects of North Korea's current situation may well be the best in the world, except possibly for those of mainland China. But Japan's underlying ability to perceive and interpret sudden developments at considerable distance from its home islands is grievously limited. Japan relied, for example, almost entirely on American intelligence for a detailed understanding of the 1990 Iraqi invasion of Kuwait, even though it had major business interests (particularly those of the Arabian Oil Company) in the Neutral Zone between Iraq and Kuwait, which lay virtually astride the invasion route.

Formally coordinating assessment of political-military intelligence for the Japanese prime minister is the Cabinet Information Research Office (Naikaku Jōhō Chōsa Shitsu), which gathers intelligence information from all the collection agencies enumerated above. With only 124 staff members, Naichō, as it is called, has very modest analytical capacities, especially compared to the massive American CIA with its sixteen thousand employees. The effectiveness of Naichō is also hampered by numerous operational problems that prevent it from becoming a really effective unit. One of the most chronic, as so often in Japanese foreign policymaking, is sectionalism.

Ever since the foundation of Naichō in 1952 at the command of former prime minister Yoshida Shigeru, its director has come from the National Police Agency and its deputy director from the Ministry of Foreign Affairs. Naichō thus finds it difficult to secure the cooperation of other agencies, and even of MOFA. These other bodies tend to withhold information from Naichō and convey it to the prime minister through their own private channels. Yet Naichō, unlike the American CIA, has no authority to station its officials abroad, or to establish domestic sources of intelligence beyond the materials provided to them by other government agencies or by the media. It thus tends to become heavily reliant on informal contacts with the private sector and journalists, as well as extensive informal travel abroad by its members, utilizing the agency's substantial budget of around $200 million annually.[33]

These are highly cumbersome and labor-intensive forms of intelligence collection. Yet Naichō remains very small, with less than one hundred members. Like other intelligence agencies, it has neither the personnel nor the inclination to become global. Instead, it focuses heavily on developments within Asia, relying on long-standing private networks in Southeast Asia and China, many dating from World War II, as well as police intelligence on the Korean peninsula.

CURRENT AND FUTURE ROLE OF THE *KANTEI*

In many of the G-7 nations, especially those with presidential systems, such as France and the United States, strong central executive institutions are able to supply central direction that can inhibit interministerial conflict and encourage a proactive foreign policy orientation. As Japan has grown more important in the global economy, and as its rising investment stakes abroad have broadened its foreign policy concerns, pressures have risen both inside and outside Japan for a more decisive top-level diplomacy. The annual G-7 summits and the proliferation of heads-of-government meetings like the 1995 Osaka APEC summit are also increasing both the opportunity and the political need for Japanese prime ministers to act more vigorously on the international stage, especially on trade and international monetary questions.

What institutional capacities does the Japanese Prime Minister's Office (*kantei*) have to act on foreign-policy issues? The accent must inevitably be placed on its limitations. The *kantei*'s weaknesses with respect to intelligence gathering and analysis have already been noted. An even more general problem exists with respect to staff. The prime minister's direct staff, consisting of private secretaries from four major government agencies (Foreign Affairs, MOF, MITI, and the National Police Agency), a single personal secretary, a chief cabinet secretary, two deputies, and a small number of others, totals only eleven. It occupies just one large room next to the prime minister's own office, in an overcrowded official residence built in 1929. This meager coterie compares to around 325 on the White House staff (not including the National Security Council, the Office of Management and Budget, USTR, and other such affiliated agencies of the Executive Office of the President), and a British prime minister's office at 10 Downing Street with a total staff of about seventy.[34]

Even the Japanese Cabinet Secretariat, located in a five-story building across the street from the prime minister's residence, but performing important support functions, has a staff of only 175, or less than half the size of Britain's Cabinet Office, which has a total staff of about four hundred.[35] For nearly a decade there have been plans to modernize the Japanese prime minister's residence and to consolidate its staff. The proposed new prime minister's residence, five floors above ground and two underground, would house a situation room and command center staffed around the clock, as well as a heliport on the roof to facilitate the movements of the prime minister. But currently the *kantei* lacks any such facility, and the schedule for its realization remains unclear.

A second limitation on the Japanese prime minister's foreign policy capacities is his staff's inability to serve as a tool for penetrating the bureaucracy, in contrast to patterns in several Western nations. Although he appoints the ministers and parliamentary vice ministers at such agencies as MOFA, MITI, and MOF, their influence over their bureaucratic subordinates is limited. Conversely, almost all of the prime minister's own personal staff, which follows him throughout virtually every day, is on loan from the ministries. Only his chief secretary (often a son or other relative) is generally unambiguously loyal. This pattern contrasts sharply to both the United States and Germany, where the chief executive has a large, independent personal or institutional staff, and to France, where the president and prime minister can both control appointments from the bureaucracy to their own personal staff.

There has, to be sure, been some modest recent enhancement in the institutional capacities of Japan's *kantei* to conduct foreign policy. Two important structural changes were made in 1986 by Nakasone Yasuhiro, prime minister from 1982 to 1987, who strongly believed that a strengthened *kantei* would be crucial to expanding Japan's role in global affairs. First of all, Nakasone established an independent Cabinet Councillors' Office on External Affairs (*naikaku gaisei shingi shitsu*). This body was made formally responsible for "providing overall coordination on important items engaging the cabinet, and for providing the overall coordination needed to preserve the unity related to policies of each administrative party, with particular emphasis on items related to foreign relations" (Cabinet Law).

The director of the office is drawn on secondment from the Ministry of Foreign Affairs; the first incumbent under Nakasone was Kunihiro Michihiko, a longtime trade negotiator with the United States, who after his term at the *kantei* was posted as Japanese ambassador to Indonesia. Its staff is drawn on secondment principally from the major ministries, including the Economic Planning Agency, MITI, MOF, MOFA, and the Ministry of Agriculture. Although this new office has not overcome the endemic problem of sectionalism that continually plagues Japanese policymaking, it does seem to have deepened MOFA input into prime ministerial decision-making. Significantly, it has given MOFA an alternate channel for liaison with the prime minister to the Cabinet Information Research Office, which is less dominated by the National Police Agency than is Naichō.

Nakasone's second structural innovation at the *kantei* was establishing the

Cabinet Security Affairs Office (*naikaku anzen hoshō shitsu*), created in July 1986 to staff the new Security Council of Japan. This latter body, including the MOFA and MOF ministers, the directors-general of the JDA and Economic Planning Agency, the chief cabinet secretary, and the head of the National Public Safety Commission, now meets under the direction of the prime minister to make major national-security-related crisis decisions, after the pattern, at least in form, of the U.S. National Security Council. Its staff is part of the Cabinet Secretariat.

An expanding White House staff was, of course, a central element in the decisive global foreign-policy role of the United States during World War II and its turbulent aftermath.[36] But the new institutional changes introduced by Nakasone into the Japanese *kantei* nevertheless failed to galvanize it into an analogous role, as dramatized by Japan's hesitant response to the Gulf crisis of August 1990. Prime Minister Kaifu Toshiki ended up canceling a major Middle East trip, scheduled for just ten days after the Iraqi invasion of Kuwait, that would have offered a golden opportunity for proactive Japanese diplomacy. Yet the structural changes introduced at the *kantei* during the mid-1980s do seem to have had some utility in improving noncrisis planning on issues such as the 1992–93 dispatch of peacekeeping forces to Cambodia and Mozambique, as well as some marginal added cohesion to prime ministerial decision-making in general.

The composite picture that we present of Japanese public-sector foreign policy institutions is thus a bifurcated one: powerful economic policy ministries, on the one hand, contrasted to underdeveloped mechanisms for intelligence-gathering and political-military power projection. In both economic and security policy areas, sectionalism is the order of the day. Although the office of the *kantei* is slowly gaining authority and institutional strength, largely because of the proliferation of high-stakes international diplomacy involving the prime minister, and the historic initiatives of Nakasone Yasuhiro, it cannot yet suppress this chronic sectionalist tendency in Japanese policymaking. Relying on state capacity alone, Japan will most likely retain a bias, for some years to come, toward being a "reactive state" on broad matters of foreign policy.[37]

Seen in the aggregate, two aspects of Japanese institutional structure stand out: its fragmented overall configuration, and the strong command that bureaucratic agencies have over many technical details of policy formation. Taken together, these characteristics generate the outstanding traditional strengths and weaknesses of Japanese foreign policy—its slowness and hesitancy in establishing broad national consensus, especially in the security sphere, and its relative strength and precision in dealing with narrow technical questions. It is no accident that Japanese foreign policy is often meticulous in its command of detail and subtle in its incremental initiatives, while lacking in comprehensive national strategy. Its difficulties in forging a decisive set of broad-gauge APEC initiatives for the 1995 Osaka summit, coupled with its modest success in crafting some incremental institutional improvements, were a clear manifestation of these mixed, institutionally rooted strengths and weaknesses.

Japan clearly lacks the institutions to support an activist foreign policy at present. Yet there are clear forces for change in the Japanese policy process and in the broader Japanese political economy. These make it dangerous for the outside world—including the United States—to take Japan's current orientation for granted indefinitely. On the economic side, Japan's trade has shifted sharply from the United States toward Asia, with Asian trade surpassing that with the United States by increasing margins ever since 1991. More than 80 percent of Japanese foreign investment profits worldwide are now generated in Asia, while the U.S. operations of Japanese firms, although large and strategically important, are at best marginally profitable. Japanese investments in Asia have more than tripled in the past decade and now total well over $70 billion.

Japanese public opinion, as noted in Chapter 5, has also shifted significantly since the Gulf War of 1991, to a frustrated stance highly critical of existing foreign policy institutions and policies. To be sure, it remains relatively positive toward the United States, but increasingly skeptical of the underlying need for American bases in Japan, and volatile in its overall orientation. The rising political power of the domestic media, in a world of fluid coalition politics, suggests that the public could be rapidly swayed by dramatic overseas developments, such as a major crisis in Korea or the Taiwan Strait coupled with a forceful media interpretation of them.

Further compounding the uncertainty surrounding Japan's foreign policy future are the momentous changes under way in the political system. During the 1993–95 period alone, Japan had four prime ministers representing three different political coalitions, after thirty-eight years of continuous one-party dominance. Major shifts in electoral patterns also flowed from introduction of a major new lower-house electoral system in January 1994.

Ultimately Japanese politics seems likely to stabilize into a pattern approaching two competitive, relatively conservative parties alternating in power, albeit with some fringe elements remaining. This could mean substantial economic rationalization, coupled with a more forceful diplomacy less beholden to political factionalism or bureaucratic influence. The dead hand of fragmented, reactive institutions on policymaking could well be broken, although probably near this century's end at the earliest. Where then might national interest and public sentiment take Japan?

China: A Revisionist Force in Pacific Affairs?

China's policy machine presents a set of problems different from those confronting Japan. Despite its huge size, China seems to have less problem with decisiveness at the very top level of policymaking than does contemporary Japan. The problems in this sprawling land of 1.2 billion people are more with consistency and coordination. For China, to a greater extent than any other major nation in global affairs today, is governed by personal networks and expediency rather than by the rule of law.

This dominance of networks (*guanxi*) in China is rooted, as we have seen before, not simply in the agnostic cynicism of a corrupted regime in transition, but more importantly in the institutional deficiencies of a new, semicapitalist political economy struggling to be born. With the Communist Party in decay and the People's Liberation Army lacking dominant influence, there are no central institutions to aggregate interests in this vast, ever more complex nation. With the revolutionary giants Mao Zedong and Deng Xiaoping now gone, prevailing leaders cannot provide a convincing substitute. The alternative is networks—amoral, inconsistent, and sometimes indecisive, but all that China has to fill its yawning leadership vacuum.

The problems for the world of the rule of *guanxi* rather than law are great in both the economic and the security spheres. McDonald's may have a twenty-year lease on a prime location in downtown Beijing, and it may build the biggest hamburger shop in the world there. But it can still lose all in an instant to a friend of the top Chinese leadership who wants to build a shopping center in the same lucrative location. Similarly, China may assent to the Missile Technology Control Regime, but those restraints will not prevent relatives of PLA generals from shipping M-11 surface-to-surface missiles to Pakistan, or Silkworm cruise missiles to the Middle East. Law has no way of constraining connections.

A major issue for the future is how the Chinese and Japanese policy machines may interact. The relationship in recent years has been relatively smooth, in part because Japan has been so reactive and underdeveloped in the political-military sphere. The relative detachment of the Chinese and Japanese economies, with trade rather than investment being the principal linkage, has also helped, since Japanese firms have been relatively unexposed to the capricious Chinese bureaucratic treatment of foreign investment, which has intensified frictions between China and the West.

The new century could well bring intensified friction as well as growing interdependence. Japan will be more proactive, which could be disconcerting to China. And China's often arbitrary regulatory and political stances, flowing from its cumbersome, decentralized, and extra-legal policy processes, will likely be irritating to Japan. A clear, destabilizing policy gap between the giants of Asia will most likely remain, but not one so insurmountable as to render a deepening Sino-Japanese embrace impossible, should broader economic or geopolitical imperatives dictate.

The Limits of Pacific Multilateralism

The policy gap is not simply a problem of national policy "machines" and how they mesh with one another. It is also an issue of broader frameworks as well. In Europe, first NATO and then the European Union brought the western half of that old and fractious continent into a transnational embrace that has stifled the deadly ancient rivalries of German, Frenchman, and Briton. Yet nothing even remotely similar has ever occurred in Asia.

Part of the blame, no doubt, lies with the chessplayer diplomat John Foster Dulles—canny, Machiavellian author of the original "hub and spokes" concept behind the San Francisco System. For him, and for a long line of Washington policymakers thereafter, multilateral institutions only undermined the leverage of the United States, which lay precisely in the absence of alternative mediators or rule-making systems. It was convenient that Asians could not talk with one another very deeply, and even more convenient that they could not talk with the Soviet Union.

The Pacific is, of course, now seemingly in transition to a new order. APEC on the economic side and the Asian Regional Forum (ARF) in security affairs provide some modicum of nominal region-wide organization. They also generate a series of regular, ongoing high-level meetings at which top officials of the Pacific can get to know one another.

Yet it is questionable whether the existing multilateral frameworks, as currently conceived, will ever resemble their European analogues.[38] For one thing, the political systems of Asia—which range from Japanese-style democracy to Chinese-style socialism—are simply too different. European union had, as its early driving course, a coalition of Christian Democratic governments with no conceivable analogue in Asia. Asian levels of economic development are also more varied than those of Europe. And the region is already linked by various potent, informal networks—from overseas Chinese traders to Japanese general trading companies—whose comparative advantage would be undermined by formal integration.

Other limits on Pacific economic and political union are imposed by the geopolitics of the region, especially the growth of China. That growth and its imposing future prospects simultaneously complicate both security and economic frameworks for the future. Securely anchoring China in an internationalist Pacific system will be an enormous future challenge for Pacific Defense.

The larger China grows, the more tempted it may be to dictate the terms of its relationship with the region to other parties. And China's prospective terms do not, from past experience, appear likely to be based on regularized trading principles. They may be congenial to those with influential ongoing networks in China, especially overseas Chinese, but could be significantly less attractive to most everyone else.

Similarly, in security matters the growth of China could also render broad, all-inclusive Pacific multilateral organizations either impossible or meaningless. Such bodies could, of course, promote confidence-building measures or increased transparency through exchange of information. They could, for example, have reduced the chance that North Koreans would shoot down stray U.S. Army helicopters over the DMZ in the midst of peace negotiations, as happened at the end of 1994. But broad, multilateral bodies would otherwise fail to deal with the preeminent security concern of most members: how to neutralize or balance any security challenges from China itself to the broader region.

The 1994 Bogor APEC declaration established ambitious benchmarks for regional integration, including free trade among the advanced nations of the Pacific by 2010.[39] Although pressing APEC members toward free trade is a laudable goal, there is grave question as to whether, given entrenched existing interests and the ambivalence of China, such a program of radical integration is realistic. Stress on such ambitious goals and the modalities of achieving them causes policymakers to neglect several other more realistic avenues for possible progress. These neglected possibilities, considered more fully in Chapter 10, include:

1. *Sector-specific economic agreements.* Selective trade deregulation and standardization, in such areas as air travel, financial services, and telecommunications, could be mightily productive for the region as a whole.

2. *Subregional security agreements.* Mutual balanced force reductions in and around the Korean peninsula, for example, could have broad positive effects, as could a subregional agreement on resource exploitation in the South China Sea.

3. *Public goods for the region as a whole.* New energy data collection and analysis centers, such as the Pacific regional center proposed by Japan in connection with the 1995 Osaka summit, for example, are clearly needed as the world

moves closer toward an energy-short future. So are job-training centers and other such educational infrastructure.

4. *Systematic policy borrowing.* The governments of Asia and the United States, as noted throughout this chapter, have very different policy machines, leading to very different sorts of policies. They could productively discuss both ways that their institutions can be made more compatible and what they can learn from one another in terms of specific policies. Concretely, Japan, for example, has much to teach the United States about energy efficiency. The United States, conversely, has a good deal to teach Japan about telecommunications and air-travel deregulation.[40]

5. *Global issues cooperation.* People throughout the Pacific have shared interests with people around the world with respect to such truly global problems as population, health, practical technology-sharing, and environmental protection. Organized multilateral cooperation between the Pacific and the broader world on concrete, obvious-to-the-public issues could promote internationalism and help combat the rising dangers of regionalism alluded to in Chapter 8. Multilateral cooperation could productively build on the work of the U.S.–Japan Common Agenda discussions, pursued under the Bush and Clinton administrations.

The Transpacific Communications Gap

The troubling problems that America and Asia have in dealing with each other are not simply rooted in the failure of government institutions. Sadly, they also extend to the broader societies themselves. There are three major dimensions: media coverage, language education, and area studies support.

East Asia's mass media generally provide broad, detailed coverage of transpacific relations. The problem is that it is sensational, and often linked to vested domestic interests whose concerns are not transparent. It preys upon the asymmetrical dependence of Asians and Americans on their mutual economic and political ties, in order to convince Asians that even though Americans may be blustering, bullying, and irrational, as they are typically pictured, Asians should go along for the sake of a vital long-term transpacific relationship. This East Asian media approach in the end no doubt strongly stimulates Asian nationalism by rendering Americans more irrational and domineering than they probably are.

American media have a converse problem: they tend to neglect the transpacific relationship altogether. As a consequence, Americans know little about Pacific issues that are furiously debated on the other side of the ocean. Few Americans, for example, know the tragic story of poor Vincent Chin, the Chinese-American mistaken for Japanese who was beaten to death by an unemployed autoworker in a Detroit bar during the late 1980s. But the story was all over the Japanese papers and remains deep in Japan's collective consciousness.

Events in Asia sometimes do suddenly catch the attention of Americans. The 1989 Tiananmen massacre stirred them deeply, and the North Korean nuclear issue was likewise widely debated five years later. U.S.–Japan trade questions also periodically surface in our public consciousness. When events in Asia finally do catch the imagination of Americans, their response is often idealistic, but naturally oblivious of the local context in Asia itself.

Despite rapidly rising economic interdependence with Asia over the past two decades, the level of actual firsthand American experience with the region has not kept pace. We have not fought a war in Asia for more than a generation, and our peacetime military forces are gradually coming home. The generations that knew Asia so intimately through World War II, the ensuing occupation, and even Korea are quietly passing from the scene, leaving only a Vietnam War generation deeply ambivalent about Asian involvement in any case.

Inhibited by everything from a strong yen to Chinese repression, tourism and academic exchange from America to Asia has become dangerously stagnant. The number of American foreign students in Japan, for example, is only one thirty-seventh that of Japanese students in the United States.[41] Transpacific aircraft are full, and the planes are mostly American. But it is increasingly Asians, rather than Americans, who are doing the flying.

Some Western nations are doing much better than the United States in coming to terms educationally and culturally, not to mention commercially, with Asia. Australia, for example, currently has 180,000 students studying Japanese—fifty times U.S. levels on a per capita basis, and fully three times a much larger America's achievement in absolute numerical terms.[42] Many Australians begin studying Asian languages even in elementary school, thus allowing them to achieve near-native accents and to attain real proficiency by the time they reach the business or professional world. Nearly 10 percent of the students in Australian universities are foreign, and principally Asian, thus allowing for a further deepening of intercultural contacts. Not surprisingly, Australia has succeeded in raising the Asia-bound share of its total exports to over 65 percent, more than double the proportion of a generation ago.

Asian language and area studies support in the United States has been languishing for close to a decade, even as the national need for Asia specialists has steadily risen. In part this shortfall is because Asian governments and corporations have reduced their backing. Even more ominous, however, in the long run, American benefactors of Asian studies have abandoned the field almost entirely. This is dangerous not only because it cripples an overall understanding of the dynamic and vital Pacific Basin, but because it makes a balanced assessment of emerging policy issues in transpacific relations much more difficult to achieve. It abandons the field to propaganda and lobbying, precisely when a balanced view is most important.

The silence in transpacific relations—particularly the chasm of high-level attention to issues of vital long-term importance to both sides' unsettling future—is deafening. The situation has changed remarkably little in the past quarter century, even as Asia's share of the global economy has tripled, and as the troubling pathologies of the post–Cold War world have begun to emerge. What will the future hold? What are the implications for Pacific Defense?

COPING WITH THE TRANSPACIFIC FUTURE

Recognizing and responding to momentous new developments across the Pacific has never been an American strong point. Almost once a generation for a full century, this nation has found itself embroiled in major wars in the Pacific: Vietnam, Korea, two world wars, and the Spanish American War before that. In no case did any significant group see these Pacific conflicts coming.

Pearl Harbor and its military analogues across the years are by no means our only instances of national myopia. Just as striking—and in many ways more prospectively important in the end—have been our economic blind spots. Few Americans predicted the Japanese economic miracle of the 1950s and 1960s, or believed it as it began to unfold before their eyes. It took well over a decade of persistent double-digit economic growth in China after Deng Xiaoping took power before Americans began to see that something fundamental was changing. Europeans have been even more oblivious to the ongoing economic rise of Asia.

This economic shortsightedness is now, it can be hoped, in the past—at least exhaustive media and scholarly coverage has made it passé. Since the Korean nuclear crisis emerged in 1993, some sense of the very real security dangers in Asia has little by little begun to permeate American thinking. But their crucial links to Asia's persistent economic success remain sadly unexplored.

We have argued here that Asia's very growth and affluence itself lie at the heart of the emerging challenges to Pacific Defense. The deadly triangle of growth, energy shortage, and armament, in the context of fluid post–Cold War geostrategic alignments, threatens to destabilize Asia, and indeed the whole Pacific. Energy shortage, the deadly but little-known link in this equation, provokes turbulence by deepening the nuclear bias of Northeast Asia, by provoking new naval rivalries centering on emerging Chinese blue-water capacity, and by deepening tensions over offshore reserves such as those in the South China Sea.

To be sure, energy insecurity is not the only subtle new danger that policies of the future must address. Others that we have outlined include surging defense-technology transfers, particularly from Russia to China as Cold War rivalries wane; instabilities in the U.S.–Japan–China triangle as Chinese economic power and commercial attractiveness rise; and the ongoing redefinition of political relationships (principally Taiwanese and Korean) in the Northeast Asian Arc of Crisis that lies at Japan's doorstep.

In the final analysis the future of Japan and the U.S.–Japan relationship must inevitably loom largest in American calculations, as we consider the future of the Pacific over the next generation. Japan, after all, remains the economic giant of the East Asian region, with roughly half of regional GNP, and one-sixth that of the entire world. Even after fifteen years of persistent double-digit growth, China's economy by most estimates remains significantly smaller. Inflation, poor roads and railways, phones that don't work, energy shortages, a generational shift in top leadership, and profound centrifugal pressures all cloud the economic and political future of that giant, yet still backward, continental state.

Even if China continues to grow at the remarkable rates of the recent past, it should not eclipse Japan in U.S. diplomatic priorities of the coming decade. To be sure, China has nuclear weapons and is undergoing a steady, thoroughgoing military buildup. It needs to be engaged, and encouraged to behave responsibly. But its basic orientation as a nonaligned great power with a steadily growing and modernizing military is largely determined by factors beyond the scope of American influence. Japan, which stands at a historic point of political and diplomatic transition, has huge potential as either ally or adversary. And American policy—not just toward Japan bilaterally, but toward Asia more generally—can substantially influence the outcome. Indeed, stabilizing a favorable, cooperative stance toward America within Japan must be one of the highest priorities of Pacific Defense.

To a much greater extent than any other major nation in the world, Japan is already a "shadow superpower" in geostrategic terms. It has huge, sophisticated, latent military capabilities that remain underdeveloped. Applying NATO account-ing standards, Japan commits to defense less than half of the share of GNP that the United States does, and around 60 percent of China's analogous ratio.[1] There is no purely economic reason that Japan could not spend much more. Similarly, it has no nuclear weapons, although it could probably develop them, in the view of U.S. defense analysts, within several months of a decision to do so. It already has over ten tons of plutonium in stock. Japan likewise has no intercontinental ballistic missiles as such, but the H-II booster rocket, operational since 1993, could be easily adapted to launch warheads anywhere on earth. In the tactical weapons area, its version of the Tomahawk missile is arguably the most accurate in the world.

Japan's defense-industrial base, like its force configurations and overall ec-onomic scale, also provides enormous latent military potential for the future. Warfare is becoming more mobile, with technology opening the prospect of long-distance and automated conflict, potentially in space. Japan's strengths in robot-ics, microelectronics, guidance systems, new materials, and computerized communications promise to assume increasing importance. It retains the pow-erful, diverse, triple options of sharing these capacities with the United States, diffusing them to possible American adversaries such as China, or retaining them secretly at home, so as to enhance long-range domestic capabilities at the expense of other nations.

The United States has grown accustomed to taking Japan for granted in geo-strategic terms—used to assuming that Japan's "shadow power" will never be turned into harder currency. Such facile assumptions, however, can no longer be taken for granted. In the twenty-first-century world into which both nations are moving, where Asia's economy apart from Japan is likely to loom larger and that of the United States progressively smaller over time, Japan's options may well be broadening; moving away from the United States and closer to Asia could become ever more feasible for Tokyo. And Japan's inclination to proactively exercise its expanding range of options, in either "Realist" or "Gaullist" fashion, could also grow, as our study of Japan's internal "struggle for strategy" in Chapter 5 has just suggested.

All this does *not* mean that we are headed toward a "coming war with Japan," or even a serious train wreck in the next year or two, despite the high-decibel rhetoric of both nations' trade negotiators. Japan's still persistent tendency to be cautious and reactive in its actual diplomacy postpones reckoning day. But the long-range trends are troubling. Without thoroughgoing changes in U.S. policy toward Asia—structural as well as substantive—these trends will render the U.S.–Japan relationship toward the second decade of the coming century (perhaps

2005–15) by far the most dangerous since the post–Pearl Harbor decade. During that turbulent interval (1941–51), from Pearl Harbor to Pork Chop Hill, nearly 200,000 Americans lost their lives in Pacific combat.

Our prescriptions for coping with the problem of Pacific Defense boil down into ten basic precepts. They include the following:

1. Remember that both economics and security are important, and that they are linked.

The waning of the Cold War does *not* mean, as some have suggested,[2] that all policy emphasis must now be placed on economic objectives at the expense of political. To be sure, the United States and Japan have a large, persistent, irritating bilateral trade imbalance that must be addressed. Without doubt, pressure must be kept on Japan to open its markets. Yet some proposals with a limited "economics only" focus are downright dangerous when viewed against the backdrop of trends we have described.[3] In particular, the United States should realize the implicit dangers of using its forces in Japan as a direct lever to secure trade-policy concessions.

Most important, linking the presence of U.S. forces in Japan to Japanese trade behavior, even implicitly,[4] dangerously undermines the credibility of U.S. security commitments to Japan. It suggests to an increasingly skeptical and nationalistic Japanese public that the cost of having U.S. troops among them is unacceptably high, and of more benefit to America than to themselves. Such trade-security linkage plays into the hands of Gaullists such as Ishihara and Mahathir, who stress the unreliability of U.S. commitments. It also plays into the hands of Chinese geostrategists tempted to believe that manipulating U.S.–Japan tensions might be profitable. In the end, such trade-security linkage could facilitate American withdrawal from Japan and accelerate Japanese rearmament, while hastening the emergence of dangerous balance-of-power politics throughout Asia. None of these developments would be in America's interest, yet all are conceivable if the "economics only" line prevails.

While security commitments should never be used directly as a bargaining chip to elicit economic concessions from Japan or other allies, there is no doubt that credible commitments generate enormous economic advantages, especially for bankers and large-scale entrepreneurs in environments where political risk is substantial. They have a clear self-interest in a stable Asia, which should make them important elements in a support coalition for a coherent Pacific Defense framework. As credible commitments help reinforce a stable political-economic environment in Asia, the support base for a stable security framework should broaden in dynamic fashion.

2. Energy is a security priority, and it needs to be addressed now.

For more than a decade, Americans have largely ignored energy-security issues, seeing no major prospect of global oil shocks on the horizon since the early 1980s. As previous chapters have shown, however, energy shortages, steadily worsening because of rapid economic growth, are looming in East Asia. They could well be a principal catalyst for arms races, nuclear proliferation, and other dangerous forms of regional instability across the region during the coming decade. Both America and Asia must plan seriously and jointly to cope with this disturbing prospect. If there is any single underappreciated economic *and* security priority for the coming decade, it surely lies here.

The greatest imperative is clearly increasing the supply of energy produced in a politically noncontentious way within the region. To be sure, energy is fungible, and there is no purely economic reason why any nation should be apprehensive about heavy reliance on any single low-cost source. Yet there is an unavoidable geopolitical dimension, in a volatile world, to such reliance—especially for nations as dependent on imports and as limited in global military influence as those of East Asia. The more self-sufficient in energy Asia as a whole becomes, the more secure its energy-short nations will be and the less severe will be their geopolitical rivalries, both in the nearby East and South China seas and for control over lines of supply from the distant Middle East.

Most important is increasing the energy self-sufficiency of China, whose recent shift to oil importer status could become a major destabilizing factor in the overall East Asian geostrategic equation well before the end of this decade. Comprehensive trilateral U.S.–Japan–China cooperation in the development of the prospectively huge Tarim Basin reserves in Sinkiang, including the building of pipelines and other infrastructure, is one concrete initiative that should be seriously explored, although the twin obstacles of trust and commercial transparency should not be minimized. Another option is studying the feasibility of large-scale multilateral oil and liquid natural gas ventures in the Sakha/Yakutia portion of Siberia and in Sakhalin. These remote yet promising areas, which could well hold close to half of the natural gas reserves on earth, could supply much of Northeast Asia's rapidly expanding energy needs for generations to come.

There are many energy imperatives, of course, apart from oil and LNG development. Coal clearly has an important future, especially in China. There it will most likely remain the staple fuel of choice in China's myriad villages for generations to come.

The issue for policy is how to make the continued, even accelerated, large-scale use of coal in China environmentally safe for the region as a whole, when acid rain already seriously scars the forests of both Japan and South Korea. MITI's Green Plan, which has already begun providing tens of millions of dollars in grants for such projects as coal power station scrubbers to control emissions, is making a start in dealing with such problems. Clearly, expanded environmental assistance to China will be an increasingly vital aspect of energy security for both the Pacific and for the broader world. Japan must take the lead.

Alternative energy is another important dimension. Some current manifestations, such as China's huge Three Gorges Dam project to harness the Yangtze, appear to be white elephants, with highly troubling environmental implications. But hydroelectric power, including large-scale generation in Siberia and long-distance transmission to the south along superconductive power lines, should be seriously explored. So should solar and geothermal energy.

The supply of secondary energy (electricity, for example) is also a high priority—power demand is rising even more rapidly across Asia than that for primary energy forms themselves. APEC expects that Asian electric power demand outside Japan will rise roughly 8 percent annually for the next fifteen years. Electrical infrastructure will require massive investment—about $1.3 trillion (at 1992 prices) between 1992 and 2010.[5]

Energy cooperation must extend beyond the mere expansion of production capacity in its various dimensions. It should also include energy technology and conservation cooperation. The developing nations of Asia, as noted in Chapter 3, are notoriously wasteful in their use of energy. Japan, by contrast, has been making major strides in conservation since the oil shocks of the 1970s and now consumes less than half the energy per capita of the United States and less than two-thirds of Germany's level. It has much to contribute to broader understanding of energy conservation in manufacturing, just as the United States has technical insights to offer with respect to alternate energy technologies such as solar and geothermal power.

Nuclear power is obviously a prominent and, especially for Northeast Asia, economically attractive option, given the region's inevitable energy shortages and the likelihood of both rising dependency on the volatile Middle East and significantly rising future energy prices. Yet severe proliferation and environmental dangers remain, the U.S.–North Korean agreement of October 1994 notwithstanding. If controversial reprocessing and breeder-reactor programs are to proceed, as current indications in Japan, Korea, and China suggest, a powerful multilateral monitoring and rule-making body analogous to Euratom, preferably involving American participation, is absolutely essential. Strong Pacific regional support for the International Atomic Energy Agency's active involvement in curbing proliferation in Asia is also important.

3. The U.S.-Japan security framework needs special attention.

Superficially the U.S.-Japan security relationship appears very solid. In July 1994 the Japanese Socialists at last recognized the U.S.-Japan Security Treaty, after more than forty years of opposition, and virtually no one in either country now talks explicitly about abandoning it. Yet beneath the surface, highly corrosive political and economic forces are at work, as Japan's sharp response to the late-1995 Okinawa rape case suggests. If left unattended, submerged irritants could, like a dangerous cancer, provoke a serious long-term crisis over the coming decade, transforming a cooperative endeavor into "us" versus "them."

There are two central problems. Most important, the stakes of the alliance are shifting, as America's long-preeminent Cold War concern of containing the Communist world grows largely irrelevant with the collapse of the Soviet Union, even as Japan's economic interest in a security guarantee for its huge and rapidly rising investments in Asia intensifies. Within Japan itself, multinational business stakes in the security framework are rising, even as the general public, influenced by an increasingly nationalistic mass media and political world, becomes ever more skeptical of the highly visible local costs to Japan of the alliance, such as crime or financial support for U.S. bases. These often appear to be more of a nuisance than an actual benefit to Japan, especially to the grassroots Japanese public.

A second major problem is related: the formula and mechanisms for sharing costs and responsibility within the alliance. Given constitutional constraints on offshore force deployments, the principal means that Japan has traditionally employed for sharing the costs of America's East Asian military presence is support for U.S. forces resident in Japan. It has done an exemplary job in this area, from the U.S. perspective, with Japan's share of nonsalary costs for U.S. forces in Japan reaching over 70 percent in 1995, including 100 percent of DOD foreign labor costs and utility bills.[6] This support package is worth $3 billion in direct support and $700 million in indirect revenues.

The problem is that these costs incurred by the Japanese government have skyrocketed to $139,000 per U.S. soldier, more than double the support payment levels in either South Korea or Germany. Indeed, support costs have risen so high that they have begun to stir political resistance in Japan. In the fiscal 1995 budget negotiations, for example, the Socialist-led Murayama cabinet, responding to such domestic disquiet, initially proposed a level of base support inadequate to meet previously agreed bilateral targets, backing away only under U.S. pressure.

The American public and much of the U.S. Congress, focusing on dollar totals relative to Japan's massive GNP, thinks Japanese support for the transpacific alliance is insufficient; the Japanese public, focusing on the day-to-day distractions of supporting bothersome foreign bases of whose value they are increasingly skeptical, thinks it is too much. There is clearly a need to redefine the notion of burden-sharing to facilitate clear, substantial contributions to mutual security interests apart from the bases.

In redefining the concept of burden sharing, a serious, long-term-oriented U.S.–Japan security policy dialogue is vital. As Joseph Nye has proposed, such discussions need to cover at least three basic subjects: (1) bilateral issues concerning defense of the Japanese homeland; (2) regional forecasts and strategies; and (3) involvement in global peacekeeping and humanitarian missions.[7] Among the special contingencies that need to be considered, consistent with Japanese constitutional constraints, are development of regional theater missile defense systems, sea-lane security, and the U.S.–Japan–Korea relationship. Clearly stable relations between Japan and Korea will be vital to maintaining Japan's moderate defense posture, and the United States can play a central role in encouraging their development.

Ultimately the support of both the United States and Japan for U.S. bases in Japan, when local costs, operational expenses, and the wages of American troops are all considered, needs to be frozen around the 50-50 level that would suggest balanced partnership. Perhaps such a bargain could be struck through a two-part agreement involving (1) an affirmation of both nations' commitment to balanced support for U.S. forces in Japan, and (2) creation of a bilateral U.S.–Japan Partnership Fund to support broader projects of comprehensive security relevance. There is an increasing need to convince the grassroots publics of both nations, growing ever more populist as recent electoral returns have shown, that their joint security alliance stands for something of real value to them personally in the post–Cold War world.

4. The United States and Japan must consciously cultivate their political relationship and cannot simply rely on free markets to preserve it.

The importance of free markets is a ritual incantation in U.S.–Japan relations, especially on the part of U.S. trade negotiators. The general principle is clearly one that all nations should cherish from the perspective of economic efficiency. But it is inadequate as a comprehensive solution to emerging U.S.–Japan economic and especially political relations. Supplementary, proactive policy cooperation across the Pacific is desperately needed, particularly in view of the deep tensions recently stirred by the combustible mixture of bitter bilateral trade talks, reported economic espionage, and crime around U.S. bases. The U.S.–Japan agenda desperately needs broadening beyond inevitably divisive issues like autos and Okinawa.

Since the late 1980s, Japan's foreign trade has been increasingly with East

Asia, rather than across the Pacific, reflecting both heavy recent Japanese investment there (as well as rising captive imports) and rapid economic growth in the Asian region. Given the cost-competitiveness of East Asian production, especially in labor-intensive sectors like clothing and simple consumer electronics, that trend toward Asia-centric trading ties should continue and even intensify. NAFTA will generate converse pressures in the United States. The upshot is that regionalism, at least along the trade dimension, is likely to intensify on both shores of the Pacific, even with the liberalization of Asian markets.

The United States has a chronic and persistent bilateral payments deficit with Japan that recently represents well over half of each nation's overall current account imbalances. While this gaping imbalance does not directly matter in economic terms and can be substantively misleading, it is important in the overall political-economic equation, which inevitably influences the national defense calculus. America supplies defense guarantees to Japan that allow it to maintain its unusual "shadow military power" orientation. The anomaly of Japan's position has been sharply compounded over the past decade by Japan's own heavy investment in Asia, which the United States also protects in the context of its regional security role.

The dual, reinforcing trade and defense asymmetries afford Japan a "free ride" that market forces can do little to correct. To the contrary, market forces accelerate Japanese investment in a U.S.–protected post–Cold War Asia whose exports compete with U.S.–based production in both the American and Japanese markets. In doing so, such forces may actually be compounding the structural dilemma of the current Pax Americana in the Pacific. They are producing an Asia, around the periphery of potential strategic adversary China, that both erodes the American industrial base through massive trade imbalances and is defended by it. They are simultaneously creating a Japan, as the 1995 Okinawa crisis showed, that is likewise deeply ambivalent, especially at the grassroots, about the costs to local dignity and society of the status quo. This situation is politically unsustainable, over the long run, in both the United States and Japan. It is dangerous to coherent Pacific Defense.

5. U.S.-Japan joint projects can help to provide, in a mutually self-interested way, the "common equity" now needed to stabilize the existing bilateral economic and security relationship.

The current crisis of the transpacific economic and security structure is three-fold: (1) the security framework to safeguard the rapidly rising Japanese investment in Asia lacks adequate long-term credibility, creating pressures for an expanded offshore Japanese security role; (2) the current economic returns to the United States for unilaterally providing a security framework for Asia are inadequate to sustain that commitment politically, given heavy U.S. trade deficits with the region and limited American investment there relative to that of major Asian nations; (3) multilateral security mechanisms are underdeveloped and cannot easily address important long-term contingencies within the region, such as Korean reunification and Taiwan-China relations.

In this fluid current situation, rendered even more dangerous by the arms-race pressures and populist resentment outlined in previous chapters, U.S.–Japan joint projects could provide a flexible, ad hoc means of offsetting the deficiencies of the existing system and stabilizing cooperative Pacific Defense. Mutual trust and some degree of commercial transparency are obviously preconditions to the success of such ventures. Astutely chosen, however, joint projects could both help significantly to reduce U.S. trade imbalances with Asia and encourage American investment in the region, while also giving the Japanese public broader reasons to support the security framework. Large-scale U.S.–Japan projects in such areas as energy, transportation, agrobusiness, marine resources, environmental protection, and communications could also create implicit security guarantees for the capital investments of both countries in such nations as Vietnam, China, the Russian Far East, and, once outstanding nuclear issues are resolved, possibly North Korea. Such projects could in turn help to improve the environment for other forms of foreign investment.

There is no question that the potential demand for large-scale U.S.–Japan infrastructure projects is enormous. The World Bank estimates that in 1994 developing countries spent $200 billion on power systems, dams, sanitation systems, roads, and telecommunications.[8] The greatest need, it suggests, is in Asia, and rising rapidly. Indeed, the bank believes that by the year 2000 infrastructure spending will absorb 7 percent of Asia's rapidly growing regional GDP, up from 4 percent in 1994. Transport and energy will be the biggest users, although China's planned telecommunications spending is also without precedent.[9] Electrical infrastructure alone could require investments in Asia of well over $1 trillion over the next fifteen years, as previously noted.

There is already evidence from outside East Asia that U.S.–Japan joint projects could actually work. The two countries are already pursuing an East Europe Environmental Initiative, a strategic cleanup program worth $1 billion. This program aims to remedy some of the enormous environmental damage remaining in Eastern Europe as a heritage of the Cold War. Practical experience gained from U.S.–Japan collaboration there could later be applied elsewhere in the world.

U.S.–Japan joint projects could help to cement bilateral economic, political, and cultural relations at many levels of society, while also addressing fundamental, long-term Asian vulnerabilities with respect to energy, food, and other natural resources that would otherwise heighten tensions within that region. Such projects could also help to stabilize political conditions, aid economic development, and create an implicit security umbrella for foreign investment in emerging nations of Asia.

One concrete area on which bilateral projects could productively focus is mass transit, a need readily appreciated by frustrated commuters everywhere. There Japanese high-speed rail technology seems clearly suited to the emerging needs of sprawling U.S. suburban areas like the Los Angeles–San Diego and Boston-Washington urban corridors. The building of Japan's Shinkansen bullet train line in the 1960s, heavily financed by the World Bank, was a centerpiece of foreign backing for Japan's own rapid economic development. It is now time for Tokyo, in a tactful way responsive to American superpower sensibilities, to return the favor. Ultimately, joint U.S.–Japan collaboration within the United States in such areas as high-speed transport could be replicated elsewhere in the world, including possibly the eastern coastal plain of China.

U.S.–Japan joint projects could also make a valuable contribution in the national security area, both by improving the quality of weaponry and, perhaps even more important, by deepening habits of cooperation in the two nations' high-technology sectors, which could all too easily otherwise confront serious techno-frictions. The December 1994 bilateral agreement on developing advanced steel for submarines, ceramic engines for tanks, and military laser technology, for example, was a positive step forward.[10] A special emphasis of such projects should be militarily defensive technological innovations, in keeping with the spirit of Japan's constitution.

Most important, U.S.–Japan joint projects need to deal with the grassroots political and psychological dimension of U.S.–Japan relations. For it is there that the dangers of schism are greatest in the newly emerging post–Cold War world. As Chapter 8 clearly suggested, it is the general public, rather than the political leadership, that is most skeptical of international entanglement. Joint projects need to convince average Americans and Japanese—the people whose eyes glaze over at the details of diplomatic summits—that the transpacific relationship holds something of value for them personally. Given the major cultural and historical gaps between the two societies, which have had limited grassroots contact apart from the bitter tragedy of World War II, such commonalities are not obvious even in the best of times. And the mid-1990s, with populist disillusionment over economic and political leadership, crime, and economic stagnation rampant in both societies, are hardly the best of times for those trying to make internationalist arguments on their own abstract merits alone.

Simply put, the isolationist argument that foreign entanglements—in this case U.S.–Japan ties—are more trouble than they are worth needs to be discredited. And U.S.–Japan joint projects that offer something tangible at the grassroots level are an effective way to do this. But political success requirements for such projects are different on the two sides of the Pacific, making multiple ventures, with a subtly different orientation in each country, probably the best approach.

In the United States, joint projects should speak to the unsolved problems of defense conversion and industrial restructuring, with some attention also to defense technology innovation, as suggested above. Joint projects could well focus on vocational education, mass transit, and job retraining. All are issues that address grassroots frustration at the stagnant living standards and rising income inequality that have plagued much of America for a full generation. Following the oil shocks of the 1970s, Japan was remarkably adept at retraining workers in basic, energy-intensive industries for productive new jobs in more knowledge-intensive sectors. While American private enterprise has been even more dynamic in many ways at accomplishing structural economic shifts, it has done so by externalizing social costs like job retraining and shifting such burdens toward government. Given current budgetary constraints in the United States and a broad popular consensus supporting tax reduction, U.S.–Japan joint projects could provide a means for addressing these clearly unmet human needs, on a nonideological and bipartisan basis.

To be politically credible in the United States, joint projects need to go beyond tokenism. The private sectors of both nations must be involved, meaning that there must be a profit element considered and included, in some symmetrical fashion. Given problems of mutual trust and the requirements of the American system, in particular, for commercial transparency, such conditions could be difficult and time-consuming to achieve. But they are not impossible, especially among firms that have existing stakes, flowing from standing business ties, in stable transpacific interdependence.

In Japan the domestic political requirements of stable transpacific interdependence are somewhat different. Most important, they involve four things: (1) evidence of American concern for the difficult living circumstances that Japanese face as their society rapidly ages; (2) appreciation of Japan's chronic underlying vulnerabilities with respect to energy, food, and other natural resources; (3) recognition of Japan's rising status in the world, as a nation with a full one-sixth of global GNP; and (4) commercial feasibility. Joint projects can address the first, second, and fourth of these concerns, although more comprehensive measures are needed in dealing with the third.

Japan today is aging more rapidly than any other major nation on earth. Over the next decade alone the share of its population over sixty-five will increase by more than a third.[11] Just as fear of losing a job in a corporate acquisition or downsizing animates many Americans, so does the postretirement future—either their own or that of their parents—worry many Japanese.

Many Americans may have expertise of use to Japan in thinking about and dealing with this problem of aging and postretirement adjustment. This country's extensive experience with lifelong adult education, or various kinds of nursing-home services, for example, might be applicable in Japan. The extraordinary, often complementary strengths of the two nations in molecular biology, pharmaceuticals, and medical equipment could be combined in a common assault on cancer. Such an effort could yield vast benefits for the world as a whole. Yet it would also produce substantial grassroots political dividends for the transpacific alliance in Japan, where cancer cases are now proliferating in that rapidly aging society.

Japan today, like most of Northeast Asia, imports virtually all of its energy, and most of its food, apart from rice. Looking to the future, it anticipates tightening markets and possible shortages in most such commodities, intensified by rapid growth and rising affluence in Asia itself. North America, conversely, is the world's preeminent food exporter, and boasts the world's most sophisticated energy-extraction and exploration technology as well.

North America, particularly the United States, can do much to lessen the resource insecurities of Japan, and East Asia more generally. Possible energy cooperation has already been discussed. Joint biotechnological research and more operational efforts to deal with emerging shortages of food and marine products will also be important in coming years.

Commercial feasibility is important to Japanese firms, as it is to American. This reality has frequently plagued transpacific negotiations over the past decade regarding defense technology transfer, among other issues. Yet Japanese firms, especially the general trading companies (sōgō shōsha) that would be core participants in most conceivable joint project schemes, tend to have longer time horizons, a broader profitability calculus, and more sensitivity to the importance of stabilizing the Pacific trade and security framework than do American counterparts. Acceptable commercial conditions could thus be less of a prospective barrier than on the U.S. side, especially given the greater capability and probable inclination of the Japanese government to assist joint projects.

Over time, multilateral, as well as U.S.–Japan bilateral, joint projects should also be considered, both within the APEC framework and independently. Given differing comparative advantage among the nations of the region, multilateral cooperation could be a highly efficient mode of operating that would at the same time promote enhanced mutual political understanding, both among the nations of Northeast Asia and more generally.

6. Korea needs special attention.

The Korean peninsula, as we have noted throughout this book, is the linchpin of the Northeast Asian Arc of Crisis. Its northern and southern halves combined have more men under arms than either the United States or the former Soviet Union. Its nuclear crisis, despite nominal agreement between the United States and North Korea, remains in fact unresolved. And sudden, momentous change in any one of many directions—collapse of the northern regime, civil war in the North, or even violence across the DMZ—is tragically easy to imagine.

Crucially, Korea lies right next to Japan—the shadow superpower whose future geostrategic orientation is of such decisive importance to Pacific Defense. Indeed, many Japanese still endorse Meiji leader Itō Hirobumi's description of Korea as a dagger pointed at Japan's own heart. Korea's future is of immense strategic importance for Japan, but the Japanese, for reasons of history, cannot easily play an activist, unilateral role there.

Korea also lies directly adjacent to China. Sino-Korean relations have deepened rapidly—in both South and North—since 1990. The collapse of the Soviet Union intensified North Korean reliance on China, while Seoul's northern diplomacy and heavy investment forged rapidly deepening ties to Beijing.

A host of considerations—heavy, dangerous military concentrations; strategic location; and a wide range of threatening yet highly possible contingencies—thus make Korea a priority subject for American defense analysis and policymaking. And there is a coherent theme that needs to be stressed: the importance of a cooperative Korea–U.S.–Japan triangle. This triangle, if intelligently and sensitively fostered, could play a key role in coping with many of the most dangerous conceivable contingencies in the North Pacific.

Should South Korea be attacked, the function of the triangle in supporting it, parallel to the Korean War, would be clear. The United States would provide direct military assistance, and Japan would presumably assist with logistical and material support. The triangle could also assist in stabilizing the economic situation of the North, and in neutralizing North Korea's nuclear threat, as it is currently doing through the Korean Energy Development Organization (KEDO), which supplies fuel oil and ultimately nuclear reactors to North Korea.

The Korea–U.S.–Japan triangle could also be highly important in the wake of Korean reunification, whenever that should occur. The triangle could mediate the flow of reconstruction aid for the North, encouraging the flow of Japanese capital while neutralizing the tensions that might flow from narrower Korean bilateral dependence on Japan. The triangle could help to anchor Korea in a pro-Pacific orientation, and to neutralize the pull of a growing China. It might also preempt an escalating Sino-Japanese struggle over Korea, a recurring pattern of conflict that has brought misfortune to that troubled peninsula many times in the past.

One troubling prospect in the event of reunification is rising pressure within Korea for the withdrawal of American forces. Such a step might also intensify pressures, in turn, for an American strategic withdrawal from Japan. Imaginative grassroots-oriented steps could be taken by the strategic triangle, perhaps linked to broader assistance also involving the North at some stage, to ease base-related frictions in the wake of reunification, and to support a meaningful, extended American presence in Korea.

One creative step to strengthen the Korea–U.S.–Japan triangle, as mentioned elsewhere in this chapter, could be involving Korean firms or the Korean government in U.S.–Japan joint projects elsewhere in the world. North Korea is the most logical locale, although only after the nuclear issue is finally resolved. Energy and infrastructure projects in China, the Russian Far East, Vietnam, and Indonesia might be other possibilities. Systematic trilateral political consultation also needs to be deepened.

7. American business in East Asia deserves special support, from both U.S. and East Asian governments.

As has become increasingly clear across the 1990s, American business is highly competitive internationally and can fend for itself in free markets. As most U.S. business people would heatedly argue, they want and need a "level playing field," but would prefer not to live at the sufferance of any government. They feel that outside free-market centers like Hong Kong and Singapore, there are few such level playing fields in Asia.

Much can obviously be done to liberalize the business conditions that American firms, and indeed all international firms, confront in Asia. Near the top of the agenda is financial deregulation, particularly in Japan, the most heavily over-regulated major financial center of the region. Banking and financial rules have much to do with patterns of savings, investment, and interest rates that perpetuate high current-account surpluses in that nation, which is by far the largest creditor in the world. Overregulation also effectively denies important business opportunities to foreign firms in pension management, insurance, and related areas, although the situation has gradually improved since late 1994 in the wake of extended U.S.–Japan negotiations.

The Structural Impediments Initiative (SII) of the Bush administration emphasized deregulation, rationalization of the distribution system, land-policy reform, and other such structural changes in the Japanese political economy. Despite an overattraction to the benefits of massive Japanese public-works spending to the United States, this broad approach had many productive features. It could have borne even greater fruit had it been systematically pursued during the political minirevolution of 1993–94 in Japan, when the Japanese domestic political scene was unusually fluid. The "results-oriented" approach that supplanted SII under the Clinton administration did, to be sure, produce some marginal gains for U.S. firms, such as expanded sales opportunities for Motorola cellular phones. It had a useful emphasis on monitoring the implementation of agreements. Yet its benefits could have been amplified with a more foresighted, longer-term structural approach more sensitive to the changing Japanese political scene.

There are other ways, aside from revisiting the structural portion of the SII agenda, that the United States can make common cause with forces for liberalization within Japan itself. One is to back the general calls of Japanese business for deregulation, as the Clinton administration has sensibly done, while insisting on a focused deregulation agenda that highlights items, like replacement auto parts, that are of special American concern. A second step could be to back a Japanese Freedom of Information Act, proposed by former reform prime minister Hosokawa Morihiro but never implemented. Such legislation could significantly increase transparency for both Japanese consumers and foreign trading partners in dealing with the Japanese bureaucracy. A third important initiative would be to utilize the Japanese mass media—especially the market-oriented business press—more strategically to highlight the benefits of America's liberalization agenda to Japanese consumers.

Freeing market forces, to reiterate, should be the central form of U.S. governmental support for international business in Asia. Market forces, however, could lead American business anywhere in the world. Yet the stability of the East Asian *security framework* ultimately requires the vigorous presence of U.S. investment in the western Pacific. The serious danger is that the kind of free markets that seem likely in the emerging East Asia of the early twenty-first century—responsive to market signals, yet heavily shaped by personal and political relationships—may not be congenial to such investment. Both Japanese and overseas Chinese trading networks are aggressive and well established. Personal connections and favors are often important. Government policies are unclear to outsiders, and frequently discriminatory against them. American firms are often unwilling to think long-term and to assume the levels of risk required for success in unfamiliar markets where the deck often seems stacked against them.

The problem is especially acute in Japan, where the book value of American investment remains only around one-fourteenth of the reciprocal Japanese investment in the United States.[12] Years of protectionist policies and covert opposition by the Japanese private sector made it very difficult for foreigners to establish a beachhead in the 1950s and 1960s, although many at that time wanted to do so.[13] The result is a chronically unbalanced transpacific investment pattern that helps to perpetuate the persistent bilateral U.S.–Japan trade deficit, despite a more than tripling in the value of the yen against the dollar across the past quarter century.[14] Although macroeconomic identities decree that savings and investment patterns will ultimately determine current account imbalances, there is no inevitability that the Japanese and American imbalances will be primarily with each other. Yet that has been the case for a full generation, despite years of trade-policy rhetoric about "opening the Japanese market." Perverse investment asymmetries, which in turn shape trade patterns, clearly appear to be at work.

Without special policy initiatives—local government tax breaks and regulatory dispensations that explicitly attract foreign investment—this intractable problem will never be solved. Japanese will always have sharply greater stakes in the United States than vice versa. The corollary could be an American drift toward isolationist policies and a Japanese uneasiness with ritualistic assurances. This combination could, in a turbulent, unfriendly Asia dominated by the Northeast Asian Arc of Crisis, ultimately lead to Japanese rearmament and a serious breakdown of Pacific Defense.

The issue is thus in the final analysis a national security question, although one couched in economic terms. Stable transpacific relations require "mutual hostages." This means giving American business incentives to go beyond narrow regionalism, and to look beyond Mexico City, Toronto, and Buenos Aires toward Shanghai and Singapore, but also Tokyo.

Concretely, such incentives need to be created in at least three areas: taxes, deregulation, and support infrastructure. Startup costs in Asian nations—particularly Japan—are unusually high by international standards, because of high land costs, rigid, traditional distribution systems, and problems of recruiting high-quality workers to work in foreign firms. Both local governments and, where possible, the U.S. government need to take steps to help U.S. firms overcome these initial startup problems and hence neutralize the long-term risk of competing in East Asia.

8. China must be treated evenhandedly.

With the largest population and second-largest land area on earth, ample raw materials, and a four-thousand-year-old civilization that long predates ancient Greece and Rome, China clearly has a legitimate claim, to the extent that any nation does, to be ethnocentric. This Middle Kingdom may well be, within thirty years, the largest economy on earth, and a major military superpower as well; its hubris is hardly likely to go away.

Clearly it is both foolish and dangerous to depreciate China or to doubt its long-term potential. Yet that very potential could well be threatening for other nations, including U.S. Pacific allies like Japan, Korea, and Indonesia, not to mention nations like India and Vietnam, even should China not so intend. The Middle Kingdom's expansive territorial claims and pronounced self-absorption virtually guarantee that foreign anxieties will persist.

A major element in foreign apprehension about China is the pervasive uncertainty about its long-term geostrategic intentions. Some of this is inevitably rooted in the unpredictability of Chinese politics, which have been extraordinarily volatile across the past half century, and may well continue to be so. But such uncertainty is greatly intensified by the chronic lack of transparency in Chinese defense planning, weapons acquisition, and even defense budgeting. China's resource commitments to defense are made very difficult to calculate, for example, by the heavy involvement of the military in broad-ranging business activities and the difficulty for outsiders in calculating how the resulting profits are deployed. Estimates by legitimate specialists on Chinese military spending for 1995 ranged all the way from the official declared budget of around $7.5 billion to more than six times that amount.[15]

The course America must tread with China is a very delicate one—made more so by proud China's continuing sensitivity to two past centuries of foreign exploitation. A major element must be mutual confidence-building: persistently consulting with Chinese leadership and building intricate transpacific networks with that vast, varied, and highly personalistic society.

One important means of confidence building must surely be deepening the U.S.–China dialogue over economic matters of real political and security significance for China, where transpacific cooperation is feasible and mutually beneficial. High on such an agenda must surely be discussions over energy, agriculture, environment, and marine resources. These are long-term concerns not only for populous China, with fourteen million new mouths a year to feed, but for all China's Pacific neighbors as well. They are also areas of notable American competitiveness and technical sophistication, where cooperation with China could bring major practical dividends, not to mention goodwill, on both sides.

Another vital element of that bridge-building to China must be a deepened dialogue at many levels with its local and regional leadership. Such a dialogue would help to monitor the deepening process of decentralization within China, described in Chapter 6, and help to minimize the risks of chaos. The emergence of more influential and internationalized regions in China, within the context of a continued formal, stabilizing national unity, may be the key to assuring that China does not become more dangerous to the broader world as it grows richer and stronger.[16]

Taiwan is a special, excruciatingly difficult case. It is becoming a full-fledged democracy, and buys twice as much from the United States as China does. Its future could bear significantly on the geostrategic orientation of Japan next door. The United States needs to thread a delicate middle road, retaining observance of the Shanghai communiqué recognizing China's unity, while also emphasizing American concern for a peaceful resolution of Taiwan's future. Building more cooperative relations with China in the energy and natural resource areas, as suggested above, should ease Chinese acceptance of inevitable American concern for Taiwan's future.

In building bridges to China, the United States must be very careful not to neglect Japan—itself a highly personalistic, face-oriented society sensitive to even subtle slights. Japan, after all, is a consistent American ally of more than forty years' standing, with immense latent military, as well as economic, potential for the future. Unlike China, it is probably too crowded and exposed geographic-ally to become a full geostrategic superpower in its own right. Yet it could easily hold the balance, farther into the twenty-first century, between China and the United States. It is some prospective superpower's natural primary strategic part-ner, and the United States would be foolish not to avail itself of such a valuable alliance possibility. If this is done by simply perpetuating existing U.S.–Japan ties, it should not unduly antagonize China. But if the U.S. alienates Japan now and tries to reverse course later as China's power becomes more apparent, such an erratic shift could damage Sino-American relations, not to mention Sino-Japanese relations, much more seriously in the long run.

There are dangerous natural tendencies in the emerging East Asian regional system, as noted in Chapter 7, toward competitive balance-of-power politics. Most dangerous of all is the not unlikely long-run prospect of a Sino-Japanese arms race. Such a development could all too easily become a major, destabilizing ele-ment of global politics a generation hence if the United States, China, and Japan are not all sensitive in their mutual relations with one another. Differently put, stabilizing the U.S.–Japan–China triangle is, as Henry Kissinger suggests, a long-run task that each of the three parties can abandon only at great risk.[17]

The way to stabilize this vital Pacific triangle, however, is emphatically not for the United States to shift deftly from one side to the other between China and Japan, in the tradition of the nineteenth-century European balance of power. This approach would, as suggested, seriously alienate Japan, a vital ally that is not good at balance-of-power politics and dislikes them intensely. Far more astute is the systematic promotion of common, mutually beneficial endeavors among the three nations in such areas as energy, food, fisheries, and environmental protection, coupled with careful American attention and fidelity to the U.S.–Japan security alliance. Such endeavors can deepen a sense of common purpose in addressing one of the greatest ultimate objectives of Pacific Defense: neutralizing resource insecurities that could otherwise destabilize East Asian relations with the broader world, as Pearl Harbor showed dramatically in 1941.

9. Multilateral frameworks are important, but only a meaningful solution for limited long-range problems.

APEC and its security analogue, the Asian Regional Forum (ARF), are positive developments that can usefully deepen transpacific interpersonal networks and build mutual confidence among their member nations, which include China. These new multilateral bodies may be constructive in coping with some important sectoral issues, such as energy and telecommunications development, as well as inspiring coordinated deregulation. But such all-inclusive bodies can easily be paralyzed by controversy. They are intrinsically incapable of dealing with conventional bilateral security confrontations or the prospect of domestic instability, other than by ceaselessly, yet often innocuously, urging additional confidence-building measures. Such intractable problems as domestic turmoil, sadly, loom large in East Asia's future, especially in the northeastern part of the region.

In the short run, small-scale, subregional cooperation, particularly that oriented toward creating free-trade-oriented "growth poles" like Hong Kong/Guangdong and the SIJORI area linking Singapore, the Johore region of southern Malaya, and the Indonesian Riau Archipelago, may be most efficient on the economic side. In security matters, subregional agreements to cope with prospective developments in Korea ("two plus four proposals" involving the two Koreas as well as neighboring powers) and the South China Sea seem very much in order.

Faced with competition from Japanese trading companies and overseas Chinese family firms, a transparent Pacific trade structure is clearly in American interests. But it will take time to achieve. General declarations of resolve to create a free-trade zone in the Pacific by 2010 or 2020 can be a useful inspiration to liberal trade forces. Yet they do not address the thorniest and most crucial long-term issues regarding the region's future. Those lie in the security sphere, although many of the most promising tools for coping with them are economic.

The heart of the matter is China. If it continues to grow and systematically arm, how can it be contained? If it does not grow, and slides instead into economic and political chaos, how can the region cope with the effects thereof? Mercifully the real policy challenges are still years off. One must hope that realistic solutions, perhaps informally and diplomatically building on China's deepening regional pluralism, will be found before then.

What we can and must do now is address the soluble aspects of the security dilemmas that China poses, as rapidly and efficiently as possible. This requires a dual strategy: extending a helping hand where that is feasible, while at the same time clearly upholding the standards of the international community where those are substantively important.

Energy in all its manifestations, including environment, power generation, conservation, and transportation, as well as resource extraction, must be at the heart of the cooperative strategy. Aiding China in any of those areas can have only positive strategic consequences, for it is China's emergence as a major oil importer that poses some of the most important prospective security challenges. China, as we have noted, has huge energy resources, but many of them are either polluting, such as coal, or inconveniently distributed across the country, such as oil and hydroelectric power.

Energy aid to China is urgent. Chinese economic growth of late has been so explosive that major bottlenecks in electric power, resource development, and transportation are rapidly developing. Those bottlenecks are threatening to constrain future growth at precisely the most politically delicate moment: that of generational transition in leadership. It is precisely now, when nationalistic forces can most easily gain ascendancy, that internationalists in China need a cooperative, helping hand in the historic task of raising living standards for nearly a quarter of the world's people and diverting them from a chauvinistic path.

It is crucial that this cooperative dimension of relations with China be multilateral, for three reasons. First, such steps could help build on to the transnational institutions like APEC that, in the long run, will be essential to anchoring such a huge, proud nation as China stably in the international system. Second, a multilateral approach will allow for systematic international support for Chinese development, keeping national rivalries among aid providers to a minimum. Third, a multilateral approach will diffuse risk, increasing the feasibility of what are inevitably large, capital-intensive projects in a politically uncertain environment.

Cooperative programs need to be accompanied by straightforward insistence on internationally accepted standards of behavior in such areas as protection of intellectual property and treatment of foreign investment. Again, the multilateral approach, where feasible, prevents any party from manipulating others in balance-of-power fashion. It also conveys firm signals that will both preserve international economic interests with China and prevent dangerous misperception of international resolve in the security sphere.

A final imperative, easily pursued, that can help to defuse long-run security dilemmas with China is promoting the broadest possible membership in Pacific multilateral organizations. Now that China is generally participating, Vietnam and, where appropriate, Mongolia, North Korea, Taiwan, Myanmar, Laos, Cambodia, and possibly India should also be included. The vigorous growth of all these nations on the periphery of China promotes national autonomy and prosperity on their parts. Membership also helps them become healthy buffers rather than weak and isolated objects of great power expansion.

10. U.S. institutions for understanding and responding to East Asia—both public and private—need major overhaul.

This needs to start with the federal government and how it is configured to deal with the challenges that East Asia presents. The most important problem is that Washington has no effective mechanism for understanding and responding to the interdependence of economics and security. Yet that is exactly what generates East Asia's most profound, if least recognized, policy challenges in the post–Cold War world.

The 1993 creation of a White House National Economic Council to complement the existing National Security Council initially seemed to be a promising start at grasping these interrelationships. But this new body had virtually no secretariat, which meant weak transmission and implementation of decisions. Furthermore, the NEC's very limited staff had virtually no expertise on security matters, just as the NSC staff has generally lacked training or experience in economics. To make matters worse, the NEC has lacked sufficient East Asian regional staff expertise and has been continually distracted in its international operations by domestic responsibilities. Yet it has nevertheless been forced to deal with a broad range of major Pacific issues, including the U.S.–Japan Framework talks, auto negotiations, and delicate trade issues with China.

Behind the gaps in the new organizational structure, as in its predecessors, is an implicit failure of conception: a belief that policy issues are all either pure security or pure economics. Some of the most important and vexing questions confronting East Asia and America, as we have persistently argued, lie precisely at the interface. The end of the Cold War does *not* mean that "security is dead." But neither does it mean that the economic origins and consequences of security concerns are irrelevant to Pacific Defense. The volatile situation on the Korean peninsula and the deepening conflict in the South China Sea are two clear cases in point.

The most urgent problem is the need to integrate economic changes into a conventional security calculus: to understand, for example, how economic growth or changing energy demand generates new patterns of military competition, or domestic political instability that can have security consequences. To deal with this new set of policy issues, the most urgent need for change is at the bodies traditionally dealing with national security: the Pentagon, the State Department, the National Security Council, and the Central Intelligence Agency. They need to be systematically organized from the transpacific relations standpoint, with two objectives in mind: (1) to relate economic change—from energy price increases to major technical breakthroughs like the advent of mobile telephones or high-temperature superconductivity—to analysis and operational capabilities; and (2) to integrate economic options, such as those relating to energy, into the national security process.

The easiest, and perhaps the most important, change to make is to give the National Security Advisor a second deputy, with an exclusive brief for economic affairs. This individual would systematically present the economic considerations often ignored in conventional, narrowly focused political-military analysis, thus forcing analytic integration at the top levels of the National Security Council. Such an innovation would be far superior to the unstructured competition between NSC and NEC prevailing after the NEC's creation in January 1993. This process has forced complex economics and security tradeoffs, such as those relating to Japan policy, sporadically and haphazardly upward into the hands of the President, who typically has neither the issue-specific background nor the time to consider them adequately.

Another innovation could be stressing the intellectual side of intelligence collection and analysis, including long-term political-economic aspects beyond the normal concerns of the intelligence community. With rapid advances in satellite and telecommunications technology, technically generated signals intelligence (SIGINT) has taken great priority in recent years. SIGINT can indeed be important in assessing short-term developments, such as new construction at the Yongbyon North Korean nuclear center. Yet human insights, including the traditions of scholarly inquiry as well as those of conventional intelligence, are urgently needed in assessing the gradual developments at the heart of most important Asian security questions, including the likely future profiles of Japan, China, and Korea.

Effectively configuring the American technology policy process for a rapidly changing transpacific future deserves special priority. One aspect is consistent government support for public efforts to stay up to date on rising Asian, especially Japanese, technological capabilities, such as the National Institute for Science and Technology (NIST) and the new Library of Congress electronic retrieval system for Japanese technical information. The Asian programs of the National Science Foundation and the National Academy of Sciences also serve important national purposes that merit public support.

With the Cold War approaching an end, even in Asia, the United States and its allies have quite appropriately moved to transform the old COCOM[18] structure of strategic technology controls. Yet the logic of this volume suggests clearly that important geostrategic uncertainties remain in Asia. This is true even as the technological level of key American allies like Japan rises to global state-of-the-art in such sensitive areas as optoelectronics—the common science of guidance systems for both industrial robots and missiles. The issue of how to promote vigorous civilian technology flows throughout the region without intensifying the threat of a local arms race requires careful study by the United States and its closest allies.

Another high priority is reorganizing the way America deals with international energy issues. The Energy Department as currently constituted tragically neglects such questions in its preoccupation with nuclear weapons manufacture. DOE supervises every aspect of this sensitive process, from the enriching of uranium to the delivery of finished weapons to the Pentagon. That quintessential Cold War responsibility has deeply shaped its consciousness, while traditionally consuming more than 70 percent of DOE's total budget.

The department has in recent years devoted some attention to domestic utility regulations, especially of nuclear power plants. Yet it has found venturing much further from its central nuclear weapons responsibilities most difficult. In an era when international cooperative mechanisms for encouraging new Pacific energy development projects are urgently needed, the Energy Department has few efficient vehicles for conducting international negotiations. Nor has it means for coordinating with other agencies, such as the State and Commerce departments, that do. DOE's counterparts as energy policymakers across the Pacific, beginning with Japan's MITI, are much more powerful and effective bodies.

Clearly the United States does not need more bloated bureaucracy for its own sake. The private sector is the heart of its competitiveness and dynamism, and recent cutbacks in Energy Department staff and funding, especially on the nuclear side, were long overdue. Yet systematic public-sector thinking and planning about long-range Pacific energy trends is fundamental to our national security, as well as our economic well-being. Some government entity capable of doing this integrated analysis and acting systematically with our allies to assure expanded, environmentally acceptable Pacific energy supplies must remain.

Another area where radical surgery is needed is at the Commerce Department. Like DOE, Commerce is massively oversized, and it includes so many unrelated agencies that it lacks a coherent basic mission. There is no good reason why the Census Bureau, the Weather Bureau, the Bureau of Economic Analysis, the Foreign and Commercial Service, and the National Aquarium should all be lumped into a single agency. In reorganizing the Commerce Department, however, Americans must not forget the powerful trade ministries that their government daily confronts abroad—not least in East Asia. Wherever the core business-support functions, especially for small-business exporters, are ultimately located, they definitely need to be preserved.

The region-specific structure of the U.S. government with respect to East Asia badly needs to be augmented, beginning with its treatment of Japan. As principal global economic competitor of the United States in most high-technology sectors, with a full sixth of global GNP and powerful latent military capacities, Japan can be either America's foremost ally or one of its principal post–Cold War antagonists, on a ten-year time frame. Over twenty years, its orientation in the U.S.–Japan–China triangle can also have fateful significance.

Given the importance of East Asia, particularly Japan, to long-term American national interests, some mechanism for systematically introducing a concern for long-term Pacific issues persistently, at high levels of any administration, is long overdue. Such a mechanism would be especially important in U.S.–East Asian relations, because the local decision-making processes of the region are often so slow and personalistic. Like a huge battleship or supertanker, East Asian nations typically change their course with little more than glacial speed. Other nations dealing with them must plan far in advance and pursue their objectives with great persistence and consistency.

One way to generate the persistent, high-level attention necessary to success in dealing with East Asia might be to designate a trusted, senior White House official—by no means necessarily a specialist—with a specific "watching brief" on U.S.–Japan or other key transpacific alliance relationships.[19] This adviser, tasked explicitly to monitor information flows rather than attempt policy coordination, could serve as a natural contact point for desk- or embassy-level officials, and could conversely encourage broader priorities not relating to country specialties. The post would serve as a bridge between area specialists and policy generalists, who desperately need better means of communication with respect to transpacific relations than they have at present.

Some will object that designating a senior official to deal with U.S.–Japan relations in a persistent way will lead to "clientelism" and a "sellout" of American interests. To be sure, the United States must militantly guard national interests, including economic priorities, and rectify many pointless, asymmetric benefits that Japan has long enjoyed. But a credible, high-level U.S.–Japan facilitator is desperately needed for three reasons.

Most important, the emerging U.S.–Japan relationship has complex, intertwined economic and security components that must be treated in a unified fashion. A single facilitator, experienced in both areas, is best equipped to do this. Second, there is a desperate need for concrete cooperative action—in the joint projects area, for example. A clear-cut coordinating mechanism is needed for this. Finally, Japanese policymaking is highly personalistic and responds most efficiently to an American structure that has a clearly designated contact point for policy communication.

In addition to enhancing White House ability to coordinate Japan policy, especially to integrate its economic and security elements, it is also crucial to intensify senior-level consultations with Japan, and indeed with East Asia more generally. Precisely because East Asian societies are so personalistic, there is much to be gained by intensified, culturally sensitive, high-level policy dialogue with them, especially since both their economic and their geopolitical importance to the United States are steadily rising. Apart from periodic, regularly scheduled presidential and cabinet-level visits to East Asia, and to APEC summits, this goal of consultation could be served by formally designating a small number of ambassadorial-rank diplomats as special counselors to the Secretary of State. These individuals, many of them fluent in East Asian languages and with extensive personal networks in East Asia, could serve as low-profile emissaries in resolving delicate conflicts. They could also provide informal advance understanding of major U.S. initiatives to slow-moving Asian political processes where appropriate, and strategically present American views to local mass media.

In the interdependent, rapidly changing global economy now emerging in the information age, America needs to think not only about "reinventing government," but about forging more effective business-government networks, both at home and abroad. One element of this is more effective and extensive Asian language education on the Australian early immersion model, as discussed in Chapter 9. Another step could be the more extensive use of universities as a forum for considering long-term policy issues. They have a detachment from the passions and the deadlines of day-to-day policymaking that can make them a useful intellectual complement to government itself, albeit necessarily one of independent judgment. In an era of budget cutbacks, they can also provide policy-relevant information and ideas in a cost-effective manner, from a public perspective.

More thought should also be given to systematically fostering interpersonal networks in Asian capitals—from Tokyo to Singapore—that better allow U.S. business and government to promote American interests. One element of this could be encouraging U.S. private-sector lobbying in Asia, to offset the extensive Asian lobbying in Washington. More systematic information exchange between

Washington officials and local American Chambers of Commerce in Asia should also be encouraged, coupled with U.S. government support for greater legitimate involvement by American business in local policy processes. The private-sector U.S.–Japan Trade Facilitation Council is a model that perhaps should be more generally applied.

U.S. private-sector representation on Asian governmental advisory boards could also be positive. Such networking and informal influence on the generation of Asian government policy proposals can be especially important in improving the terms of expanded American investment in Asia. That will ultimately be crucial in assuring both a vigorous American economic role in Asia and a stable U.S. security presence throughout the region. Both are crucial pillars of Pacific Defense.

A half century distant now from the Pacific war, the passions of that tragic struggle have cooled, and the memories have faded. Yet history still threatens, despite the complex bonds of vastly deepened economic interdependence, to repeat itself. The internal rivalries of Asia are growing ever more heated, and the shores of the broad Pacific are growing more distant from one another. It is in our hands—it is this generation's solemn responsibility—to bring them closer again.

NOTES

CHAPTER 1: ARMS RACE ASIA?

1. International Institute for Strategic Studies (IISS), *The Military Balance*, 1994–95 ed. (London: Brassey's 1994), p. 22.

2. *Ibid.*, p. 284.

3. *Mainichi Daily News*, March 6, 1995.

4. On these developments, see International Institute for Strategic Studies (IISS), *Strategic Survey*, 1993–94 ed. (London: Brassey's, 1994), pp. 41–50.

5. Estimates of the 1995 Chinese defense budget range from $7.5 billion (Chinese government) to $22 billion (International Institute of Strategic Studies), $38 billion (U.S. Defense Department), and $50 billion (U.S. Arms Control and Disarmament Agency). See *Far Eastern Economic Review*, April 13, 1995, p. 25.

6. *Far Eastern Economic Review*, October 14, 1994, p. 58.

7. U.S. Department of Energy Information Administration, *International Energy Outlook*, 1994 ed. (Washington, D.C.: U.S. Government Printing Office, 1994), p. 37.

8. International Energy Agency 1995 forecasts were for 95 million barrels/day global consumption in the year 2010. See *Wall Street Journal*, April 25, 1995.

9. Robert A. Manning and Paula Stern, "The Myth of a Pacific Community," *Foreign Affairs*, November/December 1994, p. 81.

10. "Greater China," in this usage, includes the Chinese mainland, Taiwan, Hong Kong, and Macao—areas generally recognized as Chinese, or reverting to China by the year 2000.

11. On the concept of an Islamic-Confucian connection, see Samuel P. Huntington, "The Clash of Civilizations?,"

Foreign Affairs, Summer 1993, especially pp. 45–48.

CHAPTER 2: THE NORTHEAST ASIAN ARC OF CRISIS

1. Alvin D. Coox, *Nomonhan: Japan Against Russia, 1939* (Stanford: Stanford University Press, 1985).

2. Combined forces of the fifteen members of NATO in 1990 totaled just over 4.3 million. See IISS, *The Military Balance*, 1993–94 ed., p. 33.

3. *Straits Times*, July 11, 1994.

4. Ministry of Finance data, on accumulated value of approvals and notifications basis. See Keizai Kōhō Center, *Japan 1995: An International Comparison*, p. 55.

5. Japanese Defense Agency, *Defense of Japan*, 1994 ed. (Tokyo: Japan Times, 1994), p. 46.

6. IISS, *The Military Balance*, 1994–95 ed., pp. 178–80.

7. *Ibid.*, 1993–94 ed., pp. 224, 226.

8. On the rationale for a North Korean nuclear program, see Andrew Mack, ed., *Asian Flashpoint: Security and the Korean Peninsula* (St. Leonards, Australia: Allen & Unwin, 1993).

9. *New York Times*, December 26, 1993, pp. 1, 8.

10. *Ibid.*, p. 8.

11. See, for example, Yossef Bodansky, *Crisis in Korea* (New York: Shapolsky Publishers, 1994), pp. 110–23.

12. Leonard S. Spector and Mark G. McDonough with Evan S. Medeiros, *Tracking Nuclear Proliferation* (Washington, D.C.: Carnegie Endowment, 1995), p. 103.

13. *New York Times*, December 17, 1994.

14. Spector, McDonough, and Medeiros, *Tracking Nuclear Proliferation*, p. 103.

15. Bodansky, *Crisis in Korea*, p. 92.

16. *Independent*, January 15, 1994.

17. Spector, McDonough, and Medeiros, *Tracking Nuclear Proliferation*, p. 105.

18. Bodansky, *Crisis in Korea*, pp. 96–109.

19. *Wall Street Journal*, May 16, 1994.

20. George T. Crane, "China and Taiwan: Not Yet 'Greater China,'" *International Affairs* 69, 4 (1993), p. 720.

21. See *New York Times*, August 1, 1995.

22. "Legislative History," P.L. 96-8 (Taiwan Relations Act), *U.S. Code Congressional and Administrative News*, June 1979, p. 661.

23. Martin L. Lasater, *U.S. Interests in the New Taiwan* (Boulder: Westview Press, 1993), p. 150.

24. The People's Republic of China's total trade in 1992 was just over $165 billion, although some significant additional trade occurred through Hong Kong. See *Far Eastern Economic Review, Asia Yearbook*, 1995 ed., pp. 14–15.

25. See the H. R. Haldeman, Richard Nixon, and Henry Kissinger memoirs on the nuclear aspect of Sino-Soviet relations in 1969–70.

26. Zhou Shunwu, *China Provincial Geography* (Beijing: Foreign Languages Press, 1992), p. 123.

27. See Maruyama Nobuo, ed., *Kyū Jyū Nendai Chugoku Chiiki Kaihatsu no Shikaku* (A view of Chinese regional development in the 1990s) (Tokyo: Azia Keizai Kenkyū Jo, 1994).

28. *Izvestia*, November 2, 1993.

29. See Michael T. Klare, "The Next Great Arms Race," *Foreign Affairs*, Summer 1993, p. 141.

30. *Vladivostok*, March 6, 1993.

31. On this development, see Anna V. Shkuropat, *The Emergence of Pacific Russia: A Prymorsky Perspective*

(Princeton: Princeton University Program on U.S.–Japan Relations, 1995).

32. See Russian Far East Regional Government Statistical Bureau cited in Russian Academy of Science Far Eastern Bureau Economic Research Institute (Rossia Kagaku Academy Kyokutō Shibu Keizai Kenkyū Jo), *Russia Kyokutō Keizai Sōran* (Economic almanac of the Russian Far East) (Tokyo: Tōyō Keizai Shinpō Sha, 1994), p. 165.

33. *Izvestia*, November 2, 1993.

34. William H. Overholt, *The Rise of China* (New York: W. W. Norton, 1993), p. 34. Data are from the International Monetary Fund.

35. *Wall Street Journal*, April 5, 1995.

36. Allen S. Whiting, *Siberian Development and East Asia: Threat or Promise?* (Stanford: Stanford University Press, 1981), pp. 91–92.

CHAPTER 3: LOOMING ENERGY INSECURITIES

1. Fereidun Fesharaki, Allen L. Clark, and Duangjai Intarapravich, "Energy Outlook to 2010," *Analysis from the East-West Center*, No. 19, April 1995, p. 2. In 1992 the Asia-Pacific region provided 11.4 percent of world oil production and supplied 4.5 percent of international oil reserves.

2. Keizai Kōhō Center, *Japan 1995*, p. 61.

3. Ichara Samuel, *The Business of the Japanese State* (Ithaca: Cornell University Press, 1987), p. 83.

4. Keizai Kōhō Center, *Japan 1995*, p. 58.

5. *Oil and Gas Journal*, May 10, 1993, p. 16.

6. Laura E. Hein, *Fueling Growth: The Energy Revolution and Economic Policy in Postwar Japan* (Cambridge, Mass.: Harvard East Asian Monographs, 1990), p. 51.

7. *Oil and Gas Journal*, May 10, 1993.

8. *Far Eastern Economic Review*, February 10, 1994, p. 23.

9. "APEC International Advisory Committee for Energy Intermediate Report," June 1, 1995, p. 21.

10. Far Eastern Economic Review, *Asia Yearbook*, 1994 ed., pp. 147–54.

11. Asian Development Bank, *Key Indicators of Developing Asian and Pacific Countries*, 1993 ed., pp. 300–1.

12. "APEC International Advisory Committee for Energy Intermediate Report," June 1, 1995, p. 21.

13. *Oil and Gas Journal*, May 10, 1993.

14. *Wall Street Journal*, November 17, 1994.

15. Asian Development Bank, *Key Indicators*, 1993 ed., pp. 132–33.

16. Pertamina internal data, 1994.

17. Tokio Kannoh, "Supply and Demand of Energy in China and Sustainable Development of the World," unpublished paper presented at the PECC Sixth Minerals and Energy Forum, Beijing, March 1994, p. 3.

18. JETRO, *White Paper on International Trade*, 1992 ed., p. 410.

19. *Oil and Gas Journal*, May 10, 1993, p. 40.

20. Far Eastern Economic Review, *Asia Yearbook*, 1994 ed., pp. 16–17.

21. Tokio Kannoh, "Supply and Demand of Energy in China," pp. 6–7.

22. Nicholas R. Lardy, *China in the World Economy* (Washington, D.C.: Institute for International Economics, 1994), pp. 18–19.

23. *Economist*, October 15, 1994.

24. *Ibid.*, October 1, 1994, p. 28.

25. *New York Times*, September 22, 1994.

26. *JAMA Forum*, March 1995, p. 15.

27. *Asahi Evening News*, November 11, 1994.

28. *JAMA Forum*, March 1995, p. 15.

29. *New York Times*, September 22, 1994.

30. *Oil and Gas Journal*, May 10, 1993, p. 40.

31. *Wall Street Journal*, October 10, 1994.

32. International Energy Agency, *World Energy Outlook*, 1994 ed. (Paris: Organization for Economic Cooperation and Development, 1994), pp. 194–95.

33. *Wall Street Journal*, April 18, 1994.

34. *Oil and Gas Journal*, April 26, 1993, p. 30.

35. Among the earliest and most insistently argued works was Selig S. Harrison, *China, Oil, and Asia: Conflict Ahead* (New York: Columbia University Press, 1977).

36. *Ibid.*, pp. 47–48.

37. *Ibid.*, p. 54.

38. *Economist*, August 15, 1992, p. 25. These countries include the OECD Pacific (Japan, Australia, and New Zealand), the Asian NIEs (South Korea, Taiwan, and Hong Kong), ASEAN, China, and South Asia (India, Bangladesh, and Pakistan).

39. International Energy Agency, *World Energy Outlook*, 1994 ed., p. 18.

40. Fereidun Fesharaki, Allen L. Clark, and Duangjai Intarapravich, eds., *Pacific Energy Outlook: Strategies and Policy Imperatives to 2010*, East-West Center Occasional Papers Energy and Mineral Series, No. 1 (Honolulu, March 1995), p. 42.

41. Michael J. Dunne, "Chasing Asian Tigers," *JAMA Forum*, March 1995, p. 8.

42. International Energy Agency, *World Energy Outlook*, 1994 ed., p. 155.

43. *Ibid.*, p. 186.

44. East-West Center internal data, February 1994.

45. *Financial Times*, May 24, 1993. As much as fourteen million tons of LNG will be produced annually from the field for a thirty-year period.

46. "APEC International Advisory Committee for Energy Intermediate Report," June 1, 1995, p. 21.

47. See, for example, Daniel Yergin and Thane Gustafson, *Russia 2010* (New York: Random House, 1993).

48. *Oil and Gas Journal*, May 3, 1993, p. 29.

49. On prospects for energy cooperation between Japan and Russia, see Kent E. Calder, *The United States, Japan, and the New Russia: Evolving Bases for Cooperation*, Princeton University Program on U.S.–Japan Relations Monograph Series, No. 5 (Princeton, 1994).

50. *Oil and Gas Journal*, March 8, 1993, p. 17.

51. International Energy Agency, *World Energy Outlook*, 1994 ed., p. 71.

52. Joseph Stanislaw and Daniel Yergin, "Oil: Reopening the Door," *Foreign Affairs* 72, 4 (September/October 1993), p. 86.

53. *Ibid.*

54. See Fesharaki et al., eds., *Pacific Energy Outlook*, p. 70. By narrower Western definitions, Chinese reserves are a much smaller but still impressive 200–300 million tons.

55. Lester R. Brown, *State of the World*, 1995 ed. (New York: Worldwatch Institute, 1995), p. 124.

56. *New York Times*, December 7, 1994.

57. *Economist*, August 15, 1992, p. 25.

58. Fesharaki et al., "Energy Outlook to 2010," p. 3.

59. IEA estimates suggest a global oil consumption of around 95 million barrels per day in the year 2010. See *Wall Street Journal*, April 25, 1995.

60. *Oil and Gas Journal*, August 28, 1995.

61. Far East Economic Review, *Asia Yearbook*, 1994 ed., p. 53.

62. International Energy Agency, *World Energy Outlook*, 1994 ed., p. 194.

63. *Oil and Gas Journal*, August 28, 1995.

CHAPTER 4: ASIA AND THE NUCLEAR THRESHOLD

1. U.S. Department of Energy Information Administration, *International Energy Outlook*, 1994 ed., p. 38.

2. Inter Press Service, July 21, 1994.

3. U.S. Department of Energy Information Administration, *International Energy Outlook*, 1994 ed., p. 39.

4. Inter Press Service, July 21, 1994.

5. *Korean Economic Daily*, January 12, 1995.

6. *Ibid.*, January 23, 1995.

7. Central News Agency, May 11, 1994.

8. *Ibid.*, pp. 196–97.

9. *Financial Times*, January 25, 1992, p. 1.

10. This method of transportation was chosen after the U.S. Congress, with oversight powers flowing from the Japanese reactors' American origins, refused Japan permission to ship the plutonium by air.

11. *Nikkei Weekly*, February 8, 1992, p. 1.

12. *Power Asia*, February 20, 1995.

13. *Financial Times*, January 25, 1992, p. 1.

14. *Nuclear News*, July 1992, p. 32. Fast breeders can convert uranium 238, which constitutes 99.3 percent of all uranium ore, into plutonium, while conventional light-water reactors mainly utilize uranium 235, which constitutes only 0.7 percent of uranium ore.

15. *Nihon Keizai Shimbun*, February 8, 1992.

16. *Financial Times*, January 25, 1992.

17. *Asahi Shimbun*, January 7, 1993; cited in *Nucleonics Week*, January 7, 1993, p. 7.

18. *Nucleonics Week*, January 7, 1993.

19. See Reuters World Service, BC cycle, April 22, 1994.

20. Inter Press Service, July 21, 1994.

21. *Nikkei Weekly*, March 13, 1995.

22. *Ibid.*, February 8, 1992, p. 1.

23. Reuters World Service, BC cycle, April 22, 1994.

24. On China's early nuclear weapons program, see John Wilson Lewis and Xue Litai, *China Builds the Bomb* (Stanford: Stanford University Press, 1988), especially pp. 35–218.

25. On the recent evolution of Chinese nuclear and strategic programs, see Deng Liqun, Ma Hong, and Wu Heng, eds., *China Today: Defense Science and Technology*, vol. 1 (Beijing: National Defense Industry Press, 1993), especially pp. 178–455.

26. William E. Burrows and Robert Windrem, *Critical Mass* (New York: Simon & Schuster, 1994), p. 192.

27. British Broadcasting Corporation, January 7, 1994.

28. The three principles stipulate that Japan will not make or deploy nuclear weapons, nor will it allow them to pass through its territory.

29. *Mainichi Shimbun*, August 1, 1994.

30. Spector, McDonough, and Medeiros, *Tracking Nuclear Proliferation*, p. 19.

31. On the policy debate over security policy during this period, see Daniel I. Okimoto, "Ideas, Intellectuals, and Institutions: National Security and the Question of Nuclear Armament in Japan," unpublished Ph.D. dissertation in political science, University of Michigan, 1978.

32. Burrows and Windrem, *Critical Mass*, pp. 403–5.

33. South Korea signed the NPT in July 1968, but did not ratify it until April 1975. See Spector, McDonough, and Medeiros, *Tracking Nuclear Proliferation*, p. 18.

34. Henry A. Kissinger, "Why America Can't Withdraw from Asia," *Washington Post*, June 15, 1993.

35. See Andrew Mack, "A Nuclear North Korea: The Choices are Narrowing," *World Policy Journal* XI, 2 (Summer 1994), pp. 27–35.

36. George T. Crane, "China and Taiwan: Not Yet 'Greater China,'" *International Affairs* 69, 4 (1993), p. 720.

37. Burrows and Windrem, *Critical Mass*, p. 405.

38. *Ibid.*

39. See, for example, Fred Ikle and Terumasa Nakanishi, "Japan's Grand Strategy," *Foreign Affairs*, Summer 1990, pp. 81–95.

40. See, for example, Hans Morgenthau, *Politics Among Nations* (New York: Alfred A. Knopf, 1978).

41. See, for example, the Kissinger interview in *Daily Yomiuri*, January 6, 1992; and Henry A. Kissinger, "Why America Can't Withdraw from Asia," *Washington Post*, June, 15, 1993. Also Kenneth Waltz, "The Emerging Structure of International Politics," *International Security* 18, 2 (Fall 1993), pp. 44–79.

42. Quoted in Tanaka Yasuinasa, *Gendai Nihonjin no Ishiki* (The consciousness of contemporary Japanese) (Tokyo: Chūō Kōron Sha, 1971), p. 103.

43. *Ibid.*, pp. 101–5.

44. S. Meyer, *The Dynamics of Nuclear Proliferation* (Chicago: University of Chicago Press, 1978), p. 172.

45. Peter Hayes, "The Republic of Korea and the Nuclear Issue," in Andrew Mack, ed., *Asian Flashpoint*, p. 52.

46. *Ibid.*

47. *New York Times*, June 22, 1994.

48. NHK Hōsō Yoron Chōsa Kenkyū Jo, ed., *Zusetsu: Sengo Yoron Shi* (A postwar history of public opinion in graphs), vol. 2 (Tokyo: Nihon Hōsō Shuppan Kyōkai, 1982), p. 171.

49. Only 10 percent believed abolition of nuclear weapons possible, while 84 percent disagreed.

50. *New York Times*, August 8, 1993, p. 7.

51. *Nihon Keizai Shimbun*, April 23, 1995.

52. On the Patriot program, see Michael W. Chinworth, *Inside Japan's Defense: Technology, Economics, and Strategy* (Washington, D.C.: Brassey's, 1993), pp. 67–95.

53. For a technical discussion of this point, see Michael Mazarr, *Missile Defenses and Asian-Pacific Security* (London: Macmillan, 1989), chap. 6.

54. *Defense News*, June 21, 1993, p. 12.

55. *Report from Japan, Inc.* (Yomiuri News Service), June 14, 1993.

56. *Nikkei Weekly*, July 18, 1992, p. 1. On June 18, 1992, for example, an engine fire during an H-2 test destroyed a test rocket and sharply set back development efforts.

57. See *Asahi Shimbun*, February 5, 1994, on the launching and its larger implications.

58. *Japan Times Weekly*, February 7–13, 1994, p. 164.

59. See *Defense News*, July 8, 1991, p. 1.

CHAPTER 5: JAPAN'S STRUGGLE FOR STRATEGY

1. Richard J. Samuels, *"Rich Nation, Strong Army": National Security and the Technological Transformation of Japan*

(Ithaca: Cornell University Press, 1994).

2. *Asahi Shimbun*, July 24, 1980; and *International Herald Tribune*, December 7, 1981.

3. On the deepening anticasualty orientation, see Edward Luttwak, "Post-Heroic Warfare," *Foreign Affairs*, May/June 1995, pp. 109–22.

4. See, for example, Koji Kobayashi, *Computers and Communications: A Vision of C and C* (Cambridge, Mass.: MIT Press, 1986).

5. This pattern has been deepening for well over a decade. See Kent E. Calder, "The Rise of Japan's Military-Industrial Base," *Asia Pacific Community*, Summer 1982, pp. 26–41.

6. Japan, unlike NATO, includes defense personnel expenditures under its welfare budget. Adjusted to NATO accounting standards, it spends around 2 percent of GNP annually on defense, while the U.S. spends roughly 4 percent. Japan's defense expenditures came to $53.8 billion for fiscal 1995, at current rates of exchange.

7. See Samuels, *"Rich Nation Strong Army,"* especially pp. 42–56.

8. On this tendency, see Michael W. Chinworth, *Inside Japanese Defense: Technology, Economics, and Strategy* (Washington, D.C.: Brassey's, 1992).

9. See Jeff Shear, *The Keys to the Kingdom* (New York: Doubleday, 1994), pp. 9–104.

10. *Ibid.*, p. 279; and *New York Times*, January 13, 1995.

11. Chinworth, *Inside Japanese Defense*, p. 95.

12. Ezra F. Vogel, *Japan as Number One* (Cambridge, Mass.: Harvard University Press, 1979).

13. On these generalizations, see the survey results presented in *Yoron Chōsa Nenkan* (Opinion research yearbook), various issues.

14. *Nihon Keizai Shimbun*, April 23, 1995.

15. In 1985, 1.1 percent "liked" Russia, 23.5 percent China, and 0.7 percent North Korea. See Jiji Press and Chūō Chōsa Sha, eds., *Nihon no Seitō to Naikaku* (Japan's parties and cabinet) (Tokyo: Jiji Tsushin Sha, 1992), p. 502; and *Jiji Yoron Chōsa Tokuhō* (Jiji special public opinion report), January 1995 ed. Respondents were asked to specify three countries that they "liked."

16. *Nihon Keizai Shimbun*, April 23, 1995.

17. *Asahi Shimbun*, July 29, 1995.

18. *New York Times*, August 6, 1995.

19. NHK September 1992 poll, *Yoron Chōsa Nenkan*, 1993 ed. p. 581.

20. *Ibid.*, p. 486.

21. *Nihon Keizai Shimbun*, April 23, 1995.

22. *Yoron Chōsa Nenkan*, 1993 ed., p. 486.

23. *Ibid.*, 1994 ed., p. 436.

24. *Mainichi Shimbun*, March 5, 1995.

25. *Ibid.*

26. *Yoron Chōsa Nenkan*, 1992 ed., p. 443.

27. See, for example, *ibid.*, p. 454. For a more general assessment, see Inoguchi Takashi, "Japan's Response to the Gulf Crisis: An Analytic Overview," *Journal of Japanese Studies* 17, 2 (1991), pp. 257–73.

28. *Yoron Chōsa Nenkan*, 1993 ed., p. 490.

29. For a readable, threefold book-length assessment, which omits only the Realist-Gaullist distinction, see Kenneth B. Pyle, *The Japanese Question: Power and Purpose in a New Era* (Washington, D.C.: AEI Press, 1992).

30. See, for example, Amaya Naohiro, "Chōnin Koku Nihon Tedai no Kurigoto" (The complaints of managing merchant nation Japan), *Bungei Shunjū*, March 1980; Amaya Naohiro, "Nichibei Jidōsha Mondai to Chōnin Kokka" (The U.S.–Japan automobile problem and the merchant state), *Bungei Shunjū*, June 1980; and Kōsaka Masataka, "Tsūshō Kokka Nihon no Unmei" (The fate of trading state Japan), *Chūō Kōron*, November 1975.

31. Some view this explicitly as being strategically beneficial to Japan. See, for example, Nagai Yonosuke, "Moratoriamu Kokka no Bōei Ron (The defense theory of a moratorium state), *Chūō Kōron*, January 1981.

32. See Takemura Masayoshi, *Chiisakutomo Kirai to Hikaru Kuni Nihon* (Japan: A small but sparkling nation) (Tokyo: Kōbunsha, January 1994).

33. See Yōichi Funabashi, "Japan and the New World Order," *Foreign Affairs*, Winter 1991/92, pp. 58–74; and Yōichi Funabashi, "The Asianization of Asia," *Foreign Affairs*, November/December 1993, pp. 75–85.

34. See Ozawa Ichirō, *Nihon Kaizō Keikaku* (A plan for reforming Japan) (Tokyo: Kōdansha, 1994); and its English translation, Ozawa Ichirō, *Blueprint for a New Japan* (Tokyo: Kōdansha International, 1994).

35. On the distinction between political and military Realists, see also Mike M. Mochizuki, "Japan's Search for Strategy," *International Security* 8, 3 (Winter 1983–84), pp. 152–75.

36. *Ibid.*, p. 168.

37. Hisahiko Okazaki, *A Grand Strategy for Japanese Defense* (Lanham, Md.: University Presses of America, 1986), pp. 76–83.

38. *Ibid.*, pp. 80–83.

39. See Ishihara Shintarō, *The Japan That Can Say No* (New York: Simon & Schuster, 1991).

40. See Edward J. Lincoln, *Japan's Unequal Trade* (Washington, D.C.: Brookings Institution, 1990), pp. 130–34.

41. *Far Eastern Economic Review*, November 24, 1994, p. 18.

42. Ishihara Shintarō and Mohammed Mahathir, *"No" to Ieru Asia* (The Asia that can say "no") (Tokyo: Kōbunsha, 1994).

43. Mohammed Mahathir and Ohmae Kenichi, *Asiajin to Nihonjin* (The Asians and the Japanese) (Tokyo, 1994).

44. *Asahi Shimbun*, July 29, 1995.

45. *Nihon Keizai Shimbun*, April 23, 1995.

46. See Tanaka Akihiko, " 'Haken, Konran, Sōgo Izon': Mittsu no Shinario" (Hegemony, chaos, and interdependence: Three scenarios), *Asuteion*, Summer 1994, pp. 76–83.

47. See Nakajima Mineo, *Gendai Chūgoku Ron* (Contemporary China) (Tokyo: Aoki Shoten, 1971); *Nihon Gaikō no Sentaku* (Japan's diplomatic choices) (Tokyo: Toyo Keiza Shinpōsha, 1978); and "Mittsu no Chūgoku Kyōson no Zushiki" (The three Chinas in Asia's new order), *This Is Yomiuri*, August 1992, pp. 91–103.

CHAPTER 6: THUNDER OUT OF CHINA

1. Traditionally known as Burma in the West, although Myanmar is the Burmese-language name.

2. *Far Eastern Economic Review*, December 22, 1994, p. 23.

3. See Nicholas R. Lardy, *China in the World Economy* (Washington, D.C.: Institute for International Economics, 1994), pp. 18–22.

4. *Ibid.*, p. 19.

5. *Wall Street Journal*, January 3, 1995.

6. Figures are for inhabitants over fifteen years of age. See International Bank for Reconstruction and Development, *Social Indicators of Development*, 1994 ed., p. 73.

7. *Ibid.*, p. 2.

8. Frank Ching, ed., *China in Transition* (Hong Kong: Review Publishing, 1994), p. 95.

9. See William H. Overholt, *The Rise of China* (New York: W. W. Norton, 1993), pp. 34–35.

10. On these parallels across the Communist-capitalist divide in East Asia, see Kent E. Calder and Roy Hofheinz, Jr., *The Eastasia Edge* (New York: Basic Books, 1982).

11. *Ibid.*, p. 96.

12. For a summary of Soviet reform efforts, see Marshall I. Goldman, *What Went Wrong with Perestroika* (New York: W. W. Norton, 1991).

13. Overholt, *Rise of China*, p. 49.

14. *China Statistical Yearbook*.

15. Paul T. Smith, "The Strategic Implications of Chinese Emigration," *Survival*, Summer 1994, p. 69.

16. Ching, ed., *China in Transition*, p. 91.

17. *Ibid.*, p. 51.

18. *Economist*, November 28, 1992.

19. *Wall Street Journal*, January 3, 1995.

20. United Press International, October 20, 1993.

21. China accounted for 63 percent of global executions in 1993. See *Far Eastern Economic Review*, June 9, 1994.

22. *Izvestia*, November 2, 1993.

23. Testimony by Director of Central Intelligence James Woolsey to the U.S. Senate Foreign Relations Committee, April 20, 1994.

24. Harry Harding, "On the Four Great Relationships: Prospects for the Future of China," *Survival*, Summer 1994, p. 36.

25. Wolfram Eberhard, "Chinese Regional Stereotypes," *Asian Survey*, no. 12, 1965.

26. On the profile of regionalism in China, see Gerald Segal, *China Changes Shape: Regionalism and Foreign Policy*, Institute for Strategic and International Studies, Adelphi Paper No. 287 (London, March 1994).

27. *Ibid.*, p. 14.

28. *Ibid.*, p. 17.

29. David Shambaugh, "The Emergence of Greater China," *China Quarterly*, 1993, p. 656.

30. Allen Cheng, "Toward a Mighty Megalopolis, " *Asia, Inc.*, October 1994.

31. *Economist*, July 23, 1994, p. 63.

32. *Economist*, August 19, 1995, p. 166.

33. George T. Crane, "China and Taiwan: Not Yet Greater China," *International Affairs* 69, 4 (1993), p. 715.

34. *Ibid.*, p. 716.

35. Harding, "On the Four Great Relationships," p. 682.

36. Shao-chuan Leng and Cheng-Yi Lin, "Political Change on Taiwan: Transition to Democracy?" *China Quarterly*, December 1993, p. 829.

37. *Ibid.*, p. 827.

38. *Far Eastern Economic Review*, August 3, 1995, p. 17.

39. *Japan Times*, July 29, 1995.

40. *International Herald Tribune*, July 26, 1995.

41. *Far Eastern Economic Review*, January 26, 1995, p. 19.

42. See *Far Eastern Economic Review*, July 27, 1995.

43. See Far Eastern Economic Review, *Asia Yearbook*, 1995 ed. (Hong Kong: Review Publishing, 1995), p. 121.

44. There were 66,000 emigrants from Hong Kong in 1992, 53,000 in 1993, and around 62,000 in 1994. See *South China*

Morning Post, October 10, 1994.

45. China's neighbors include Afghanistan, Bhutan, India, Kazakhstan, North Korea, Kyrgyzstan, Laos, Mongolia, Nepal, Pakistan, Russia, Sikkim, Tajikstan, and Vietnam. China also borders two contiguous territories, Hong Kong and Macao.

46. *New York Times*, January 2, 1995.

47. On the details, see Gerald Segal, *Defending China* (Oxford: Oxford University Press, 1985), pp. 80–230.

48. Nicholas D. Kristof, "The Rise of China," *Foreign Affairs*, November/December 1993, p. 72.

49. See, for example, Research Institute for Peace and Security, Tokyo, *Asian Security*, 1993–94 ed. (London: Brassey's, 1993), p. 97.

50. See John Caldwell, *China's Conventional Military Capabilities, 1994–2004: An Assessment* (Washington, D.C.: Center for Strategic and International Studies, 1994), p. 15.

51. *New York Times*, January 2, 1995.

52. Tass News Service, January 15, 1995.

53. *Japan Economic Newswire*, February 7, 1994.

54. Caldwell, *China's Conventional Military Capabilities*, p. 15.

55. *Ibid.*

56. On the process of their development, see John Wilson Lewis and Xue Litai, *China's Strategic Seapower: The Politics of Force Modernization in the Nuclear Age* (Stanford: Stanford University Press, 1994).

57. *Jane's Intelligence Review*, January 1, 1994.

58. *Asahi Evening News*, April 21, 1995.

59. *Japan Economic Newswire*, February 7, 1994.

60. On this concept, see Samuel P.

Huntington, "The Clash of Civilizations?" *Foreign Affairs* 72, 3 (Summer 1993).

61. Figures are for 1989–92. See Alexander T. Lennon, "Trading Guns, Not Butter," *China Business Review*, March 1994.

62. *New York Times*, January 5, 1995.

63. Reuters World Service, October 11, 1994.

64. On Chinese nuclear relations with these nations, see Spector, McDonough, and Medeiros, *Tracking Nuclear Proliferation*, pp. 49–56, 119–33.

65. Burrows and Windrem, *Critical Mass*, p. 342.

66. United Press International, November 21, 1994.

67. Burrows and Windrem, *Critical Mass*, p. 342; and *New York Times*, January 5, 1995.

68. Keizai Kōhō Center, *Japan 1995*, p. 31.

69. Pacific Economic Cooperation Council, *Pacific Economic Outlook*, 1994–95 ed., p. 22.

70. *Wall Street Journal*, March 10, 1995.

71. *Ibid.*

72. *Ibid.*

73. *Ibid.*

74. See Keizai Kōhō Center, *Japan 1994: An International Comparison*, p. 30.

75. *Wall Street Journal*, January 3, 1995.

CHAPTER 7: ASIA'S NEW BALANCE-OF-POWER GAME

1. See Chalmers Johnson and E. B. Keehn, "The Pentagon's Ossified Strategy," *Foreign Affairs*, July/August 1995, pp. 103–14.

2. See Henry A. Kissinger, *Diplomacy*

(New York: Simon & Schuster, 1994), p. 826.

3. See Sun Tzu, *The Art of War* (Boston: Shambhala Publications, 1988), pp. vii, 56.

4. See Bonnie S. Glaser, "China's Security Perceptions: Interests and Ambitions," *Asian Survey*, March 1993, pp. 252–71; and Hong Liu, "The Sino–South Korean Normalization," *Asian Survey*, November 1993, pp. 1083–1104.

5. *Asahi–Toa Ilbo* joint public opinion poll, published in *Asahi Shimbun*, July 29, 1995.

6. *Ibid.*

7. On this important issue also see Aaron L. Friedberg, "Ripe for Rivalry: Prospects for Peace in Multipolar Asia," *International Security* 18, 3 (Winter 1993/94), pp. 5–33.

8. See Samuel P. Huntington, "America's Changing Strategic Interests," *Survival* 33, 1 (January/February 1991), p. 12. Clearly many of the key players in East Asia's emerging power game, including China, Vietnam, and Indonesia, are undergoing this sort of experience, as India and a reunified Korea may well also do in the relatively near future.

9. See, for example, Harry Harding, ed., *China's Foreign Relations in the 1980s* (New Haven: Yale University Press, 1984).

10. On this pattern see Kent E. Calder, "Japanese Foreign Economic Policy: Explaining the Reactive State," *World Politics*, July 1988.

11. The interpersonal, economic, and political distances—many flowing from the combined negative heritage of the Cold War, World War II, and the preceding periods in which Japan was the primary colonial occupier—are great. Territorial disputes, including the Russo-Japanese conflict over the Northern Territories, occupied by Russia just after the end of World War II, make things even worse.

12. Aaron Friedberg, "Ripe for Rivalry," *International Security*, Winter 1993/94, p. 22.

13. On how that past is recently being interpreted, see Allen S. Whiting, *China Eyes Japan* (Berkeley: University of California Press, 1989).

14. See *Far Eastern Economic Review*, December 22, 1994, pp. 22–25.

15. In China, as indicated in Table 7-2, there was only a 23 percent increase in official government defense spending, but IISS estimates nevertheless placed the actual aggregate increase during 1991–93 alone at 46.8 percent—one of the highest rates of increase in the East Asian region.

16. IISS, *The Military Balance*, 1995–96 ed., p. 113.

17. Japan Defense Agency, *Defense of Japan*, 1991 ed. (Tokyo: Japan Times, 1991), p. 33.

18. IISS, *The Military Balance*, 1995–96 ed., p. 170.

19. *American Online*, February 20, 1995.

20. See IISS, *The Military Balance*, 1995-96 ed., pp. 182–83.

21. Andrew Mack and Desmond Ball, "The Military Buildup in Asia-Pacific," *Pacific Review* 4, 2 (1992), p. 200.

22. Joseph R. Morgan, *Porpoises Among the Whales: Small Navies in Asia and the Pacific*, East-West Center Special Reports, No. 2 (Honolulu, March 1994), p. 32.

23. *Ibid.*, p. 30.

24. Caldwell, *China's Conventional Military Capabilities*, pp. 6–10.

25. See Louise Levathes, *When China Ruled the Seas: The Treasure Fleet of the*

Dragon Throne, 1405–1433 (New York: Simon & Schuster, 1994).

26. See *Yomiuri Shimbun*, January 15, 1995.

27. Andrew Mack, "The Military Buildup in Asia-Pacific," *Pacific Review* 5, 3 (1992), p. 200.

28. See Morgan, *Porpoises Among the Whales*, pp. 18–23.

29. *Business Times*, February 2, 1994.

30. Tom Clancy, *Submarine* (New York: Berkley Books, 1993), p. 297.

31. IISS, *The Military Balance*, 1995–96 ed., p. 181.

32. See "Over the Cliff," *Aviation Week and Space Technology*, August 3, 1992, p. 19.

33. *U.S. News & World Report*, November 14, 1994.

34. British Broadcasting Corporation, November 16, 1994.

35. *Wall Street Journal*, April 5, 1995.

CHAPTER 8: ASIA AND THE TWILIGHT OF GLOBALISM

1. James Baker, "America in Asia," *Foreign Affairs*, Winter 1991–92, pp. 1–18.

2. See Far Eastern Economic Review, *Asia Yearbook*, assorted issues.

3. The Manning-Stern estimate of 45 percent seems roughly correct. See Robert A. Manning and Paula Stern, "The Myth of a Pacific Community," *Foreign Affairs*, November/December 1994, p. 83.

4. *Ibid.* In the case of Japanese exports the comparable drop was from 39 to 29 percent.

5. Far Eastern Economic Review, *Asia Yearbook*, 1987 and 1993 eds.

6. *Ibid.*

7. PECC, *Pacific Economic Outlook, 1994–1995* (Washington, D.C.: U.S.

Council for PECC, 1994), p. 22.

8. Figures from U.S. Department of Commerce, *Survey of Current Business*, assorted issues.

9. Actual fiscal 1993 sales of the nine general trading companies were around $960 billion at then-current exchange rates, of which 23.4 percent was offshore. See Keizai Kōhō Center, *Japan 1995*, p. 45.

10. U.S. Department of Commerce, *Survey of Current Business*, July 1993, p. 121.

11. Keizai Kōhō Center, *Japan 1995*, p. 55. Figures are as of March 31, 1994.

12. The total was $17.3 billion for fiscal 1994. See Jiji Press Ticker Service, June 6, 1995.

13. *Washington Post*, November 18, 1994.

14. *Wall Street Journal*, May 27, 1995.

15. Matsushita sold 80 percent of MCA to Seagram for $5.7 billion. Between 1990 and April, 1995 the yen appreciated from 144 to the dollar to around 85. See *New York Times*, April 16, 1995.

16. *Wall Street Journal*, May 22, 1995.

17. Ministry of Finance data, from Keizai Kōhō Center, *Japan 1995*, p. 55.

18. Bank of Japan, *Kokusai Hikaku Tōkei* (Comparative international statistics), 1994 ed. Cited in Keizai Kōhō Center, *Japan 1995*, p. 15.

19. *Business Week*, November 6, 1995, p. 26.

20. *Wall Street Journal*, May 15, 1995.

21. *Tokyo Business*, May 1994, p. 19.

22. *Far Eastern Economic Review*, March 18, 1994, p. 45.

23. U.S. Department of Commerce, *Survey of Current Business*, July 1993, p. 122.

24. Japanese foreign aid to Latin America, for example, reached $734 million in 1993 compared to $623 in U.S. aid to that

region. See *Wall Street Journal*, May 18, 1995.

25. Kishore Mahbutani, "The Pacific Way," *Foreign Affairs*, January/February 1995, pp. 100–101.

26. See Thomas S. Arrison, C. Fred Bergsten, Edward M. Graham, and Martha Caldwell Harris, eds., *Japan's Growing Technological Capability* (Washington, D.C.: National Academy Press, 1992).

27. Kishore Mahbutani, "The Pacific Way," p. 104.

28. Samuel P. Huntington, "The Clash of Civilizations?," pp. 25–27.

29. William Caspary, "The Mood Theory: A Study in Public Opinion and Foreign Policy," *American Political Science Review*, June 1970, pp. 536–47.

30. John A. Reilly, ed., *American Public Opinion and U.S. Foreign Policy, 1995* (Chicago: Chicago Council on Foreign Relations, 1995), p. 5. The "leaders" sample included 383 members of Congress and the Executive Branch, international vice presidents of major firms, media editors and columnists, scholars, and leaders of churches, labor unions, and special interest groups relevant to foreign policy.

31. *Ibid.*, p. 13.

32. Eugene R. Wittkopf, *Faces of Internationalism: Public Opinion and American Foreign Policy* (Durham, N.C.: Duke University Press, 1990), p. 140.

33. Reilly, ed., *American Public Opinion*, p. 14.

34. Michael Mandelbaum and William Schneider, "The New Internationalism," in Kenneth A. Oye, ed., *Eagle Entangled: U.S. Foreign Policy in A Complex World* (Boston: Little, Brown, 1979).

35. Reilly, ed., *American Public Opinion*, p. 21.

36. *Ibid.*

37. On this distinction, see Robert B. Reich, *The Work of Nations* (New York: Vintage Books, 1991).

38. Keizai Kōhō Center, *Japan 1995*, p. 18.

39. *Fortune*, February 20, 1995, p. 130.

40. In 1979, for example, the Steel Workers of America had 964,000 members, whereas in 1991 it had only 459,000. See *Statistical Abstract of the United States*, 1994 ed., p. 435.

41. In 1992, 15.8 percent were members. See *ibid.*, p. 436.

42. See *International Herald Tribune*, July 29, 1995.

43. In the 1995 CCFR poll, only 50 percent of the public, compared to 88 percent of leaders, favored giving economic aid to other nations—the widest leader–public opinion gap in the poll. See Reilly, ed., *American Public Opinion*, p. 39.

44. See especially the proposals for a National Security Restoration Act in *The Contract with America* (Washington, D.C.: Republican National Committee, 1994), pp. 92–95.

45. *Wall Street Journal*, November 30, 1994.

46. The Cato Institute, *The Cato Handbook for Congress* (Washington, D.C.: Cato Institute, 1995), p. 279.

17. It would be divisive, however: 84 percent of leaders in the 1995 CCFR poll but only 45 percent of the general public favored the use of U.S. troops if South Korea were invaded again by the North. See Reilly, ed., *American Public Opinion*, p. 39.

48. On this notion see Chalmers Johnson, "The End of the Japanese-American Alliance," *Nixon Center Program Brief* 2, 2 (February 1995).

CHAPTER 9: THE POLICY GAP

1. On overseas Chinese entrepreneurship and its organization, see S. Gordon Redding, *The Spirit of Chinese Capitalism* (New York: Walter de Gruyter, 1990).

2. On the organized Japanese private sector, see Kent E. Calder, *Strategic Capitalism: Public Policy and Private Purpose in Japanese Industrial Finance* (Princeton: Princeton University Press, 1993), especially pp. 134–73.

3. See, for example, David Halberstam, *The Reckoning* (New York: William Morrow, 1980).

4. See Robert S. McNamara, *In Retrospect* (New York: Times Books, 1995).

5. See Harry Harding, "Asia Policy to the Brink," *Foreign Policy*, Fall 1994, pp. 57–74.

6. The Commerce Department was founded in 1901. Japan's Ministry of Commerce and Industry, the precursor of MITI, was founded in 1925, while MITI itself, more or less as currently constituted, dates from 1950.

7. On the organized Japanese private sector, for example, see Calder, *Strategic Capitalism*, pp. 134–73.

8. See, for example, David Osborne and Ted Gaebler, *Reinventing Government: How Entrepreneurial Spirit Is Transforming the Private Sector* (New York: Penguin, 1993).

9. See David A. Deese, ed., *The New Politics of American Foreign Policy* (New York: St. Martin's, 1994).

10. Arthur J. Alexander, "Sources of America's Asia Policy in the Clinton Administration," *JEI Reports*, April 1995.

11. On evolving capabilities, see, for example, James Bamford, *The Puzzle Palace* (New York: Penguin, 1983).

12. See Burrows and Windrem, *Critical Mass*, pp. 426–32.

13. *Investors Business Daily*, January 5, 1995.

14. *Wall Street Journal*, December 15, 1994.

15. *New York Times*, December 17, 1994.

16. See Alvin Toffler, *The Third Wave* (New York: William Morrow, 1980).

17. For a reasonably accurate, if polemical, discussion, see Pat Choate, *Agents of Influence* (New York: Alfred A. Knopf, 1990).

18. One of the best was Harry Kern, *Newsweek*'s longtime Tokyo Bureau chief. See John Roberts, *Mitsui: Three Centuries of Japanese Business* (Tokyo: Weatherhill, 1974).

19. *Economist*, December 7, 1991, p. 38.

20. Keizai Kōhō Center, *Japan 1995*, p. 99.

21. Robert M. Orr, *The Emergence of Japan's Foreign Aid Power* (New York: Columbia University Press, 1990), p. 50.

22. Keizai Kōhō Center, *Japan 1995* and *Japan 1985*.

23. *Ibid.*

24. There were roughly 561,000 bureaucratic slots authorized in 1985, apart from those at the Ministry of Foreign Affairs and MITI. This number had shrunk to about 551,000 by 1992. See *ibid.*

25. Keizai Kōhō Center, *Japan 1995*, p. 94.

26. Japan's military is known as the Self-Defense Forces, while the bureaucratic entity that supervises it, and of which it is nominally a part, is the Japan Defense Agency.

27. IISS, *The Military Balance*, 1995–96 ed., pp. 176–92.

28. On the broader political processes by which defense spending has been inhibited in Japan, see Kent E. Calder, *Crisis and Compensation: Public Policy*

and Political Stability in Japan (Princeton: Princeton University Press, 1988), pp. 411–39.

29. On Japanese Gulf War policymaking, see, for example, Asahi Shimbun Sha, Wangan Sensō to Nihon (The Gulf War and Japan) (Tokyo: Asahi Shimbun Sha, 1991); Inoguchi Takashi, "Japan's Response to the Gulf Crisis: An Analytic Overview," Journal of Japanese Studies 17, 2, pp. 257–73; and Kent E. Calder, "Japan in 1991: Uncertain Quest for a Global Role," Asian Survey, January 1992, pp. 32–41.

30. General Affairs Agency Administrative Management Bureau (Sōmuchō Gyōsei Kanri Kyoku), Gyōsei Kikō Zu (Graphs of administrative structure), 1992 ed. (Tokyo: Gyōsei Kanri Kenkyū Center, 1992).

31. Ibid., p. 163.

32. For further description of Japanese intelligence agencies, see, for example, "On the Way to Securing a World Position? Japan's Intelligence Agencies and Their Activities," Japan Quarterly 29 (April–June 1982); and Kenji Hayao, The Japanese Prime Minister and Public Policy (Pittsburgh: University of Pittsburgh Press, 1993), pp. 173–76, 308.

33. Hayao, The Japanese Prime Minister and Public Policy, p. 175. Roughly half of the Naichō annual budget of 1.8 billion yen goes to outside groups for commissioned research.

34. See John Hart, The Presidential Branch (New York: Pergamon Press, 1987), pp. 97–101; and Colin Campbell, Governments Under Stress: Political Executives and Key Bureaucrats in Washington, London, and Ottawa (Toronto: University of Toronto Press, 1983), p. 19.

35. Hayao, The Japanese Prime Minister and Public Policy, p. 182.

36. See Bradley H. Patterson, Jr., The Ring of Power: The White House Staff and Its Expanding Role in Government (New York: Basic Books, 1988).

37. For a fuller exposition of this concept, see Kent E. Calder, "Japanese Foreign Economic Policy Formation: Explaining the Reactive State," World Politics 40, 4 (July 1988), pp. 517–41.

38. On some of the difficulties, see Robert A. Manning and Paula Stern, "The Myth of the Pacific Community," Foreign Affairs, November/December 1994, pp. 79–93.

39. See Second Report of the Eminent Persons Group, Achieving the APEC Vision: Free and Open Trade in the Pacific (Singapore: Asia-Pacific Economic Cooperation Secretariat, August 1994).

40. On such functional cooperation within the Pacific Basin, see Mark Borthwick, ed., Advancing Regional Integration (Singapore: Pacific Economic Cooperation Council, 1994), especially pp. 51–110.

41. In 1994, for example, only 1,146 Americans were studying in Japan, compared to 43,000 Japanese studying in the United States. See International Herald Tribune, October 30, 1995.

42. New York Times, August 13, 1995.

CHAPTER 10: COPING WITH THE TRANSPACIFIC FUTURE

1. See Japan Defense Agency, Defense of Japan, 1993 ed. (Tokyo: Japan Times, 1993), p. 109.

2. See, for example, Johnson and Keehn, "The Pentagon's Ossified Strategy."

3. See ibid., which suggests encouraging balance-of-power tendencies in Asia, and a possible withdrawal of U.S. troops.

4. James Fallows, Looking at the Sun: The Rise of the New East Asian Eco-

nomic and Political System (New York: Pantheon, 1994), pp. 440–41.

5. "APEC International Advisory Committee for Energy Intermediate Report" (unpublished), June 1, 1995, p. 27.

6. Office of the U.S. Secretary of Defense, Report on Allied Contributions to the Common Defense, May 1994, p. 3.

7. Joseph F. Nye, Jr., "The Case for Deep Engagement," Foreign Affairs, July/August 1995, p. 97.

8. Wall Street Journal, June 20, 1994.

9. Ibid.

10. For details, see Kyodo News Service, December 14, 1994. One prospective fruit of this cooperation would be to develop submarines able to operate at depths 20 percent greater than is possible with existing technology.

11. Currently around 14 percent, the share of Japan's population over sixty-five will likely rise to nearly 20 percent by 2005, according to Japanese Ministry of Health and Welfare projections. See Keizai Kōhō Center, Japan 1995, p. 9.

12. In 1994 the accumulated value of Japanese investment in the United States was $177.1 billion, while that of U.S. investment in Japan was $12.2 billion. See ibid., pp. 55, 57.

13. See Mark Mason, American Multinationals and Japan: The Political Economy of Japanese Capital Controls, 1899–1980 (Cambridge, Mass.: Harvard University Council on East Asian Studies, 1992), pp. 150–98.

14. On the investment imbalances and the reasons for them, see Dennis J. Encarnation, Rivals Beyond Trade: America Versus Japan in Global Competition (Ithaca: Cornell University Press, 1992).

15. Far Eastern Economic Review, April 13, 1995, p. 25.

16. On this point see Gerald Segal, "Deconstructing Foreign Relations," in Goodman and Segal, China Deconstructs, p. 352.

17. Kissinger, Diplomacy, p. 829.

18. Coordinating Committee on Multilateral Export Controls.

19. On this concept see I. M. Destler, Hideo Sato, Priscilla Clapp, and Haruhiro Fukui, Managing an Alliance: The Politics of U.S.–Japanese Relations (Washington, D.C.: Brookings Institution, 1976), p. 194.

INDEX

Bismarck, Otto von, 133
Blueprint for Building a New Japan (Nihon Kaizō Keikaku) (Ozawa), 96
Bretton Woods Agreement (1944), 150–151, 152, 153
Brezhnev, Leonid, 34
Brunei, 47, 56–57, 120
Buchanan, Pat, 169
Burma, *see* Myanmar
Bush, George, 4, 146
Bush administration, 14, 30, 71, 77, 213

C abinet Information Research Office, Japanese, 187, 188–189, 190
Cabinet Security Affairs Office, Japanese, 190–191
Cambodia, 96, 160, 185, 192, 220
Canada, 43, 53, 77, 157, 159, 160, 184
Carter, Jimmy, 76
Carter administration, 14, 71, 72, 77
Cato Institute, 169
CEA, 173
Central Intelligence Agency (CIA), 26, 91, 173, 183, 187, 188, 221
Cheney, Richard, 71
Chernobyl disaster, 65, 67
Chiang Ching-kuo, 115
Chiang Kai-shek, 17, 104, 119
Chiashan project, 4, 33
Chicago Council on Foreign Relations, 164, 165
Chin, Vincent, 196–197
China, People's Republic of, 11, 21, 23, 47, 48, 79, 81, 86, 89, 90, 91, 93, 104–126, 161, 162, 175, 177, 178, 186, 187, 188, 199, 221
    affluence in, 109–112
    air arms modernization and, 146
    air travel in, 55–56
    arms race and, 139
    auto industry in, 51, 55
    balance of power and, 128, 129, 131, 132, 133, 134, 135, 137–138, 139, 142–145
    border clashes and, 120–121
    Civil War generation of, 119
    coal energy in, 49, 58, 63, 64
    collapse of Soviet Union and, 145
    consumer revolution in, 50–51
    crime rate in, 111
    defense spending of, 2–5, 139
    Deng's reforms and, 107–108, 113, 156
    economic growth of, 105–107, 195
    emerging middle class of, 50–51
    fast-breeder reactor project of, 70–71
    foreign investment in, 106, 174
    global economy and, 108–109
    globalism and, 151, 154, 156–157
    GNP of, 38, 109, 142
    Gulf War and, 145
    Hong Kong reunification with, 117–119
    illegal immigration and, 36, 110–112
    India's conflict with, 120, 135
    Iran's arms trade with, 123–124
    Japan and emergence of, 101–102
    Japan's energy rivalry with, 137–138
    Japan's relationship with, 193–194
    Korean reunification and, 29, 127, 149, 212–213
    Korea's historical relationship with, 138
    Lee Teng-hui's U.S. visit and, 30, 32, 74, 115
    literacy rate in, 106
    Middle East arms market and, 60, 123–124, 147, 193
    military modernization and, 121–123
    military spending in, 106–107, 216
    missile production program of, 147
    Myanmar and, 104–105
    nationalism and, 120, 123
    natural gas reserves of, 56–57, 202
    naval arms race and, 141, 142–143
    naval tradition of, 143
    nuclear program of, 6, 19, 70–71
    nuclear weapons program of, 71–72, 73, 74, 76, 77, 122–124
    offshore oil reserves of, 52, 53–54, 58–59, 135, 202
    offshore territorial claims of, 8–9, 42, 49, 58, 120, 122, 135, 143–144
    oil consumption in, 49, 50, 51–52
    as oil importer, 7–9, 60–61, 219
    People's Car project of, 6, 51, 55
    policymaking machinery of, 193–194
    political decentralization in, 111–113
    population of, 111, 125
    populism in, 107–108
    regional differences in, 109
    regionalism in, 112–115
    revolutionary past of, 173–174
    Russian Far East and, 34, 36–40, 120–121